Opportunities for ENVIRONMENTAL APPLICATIONS of MARINE BIOTECHNOLOGY

Proceedings of the October 5-6, 1999, Workshop

Board on Biology
Oceans Studies Board
National Research Council

NATIONAL ACADEMY PRESS
Washington, D.C.

NATIONAL ACADEMY PRESS • 2101 Constitution Avenue, N.W. • Washington, D.C. 20418

NOTICE: The project that is the subject of this report was approved by the Governing Board of the National Research Council, whose members are drawn from the councils of the National Academy of Sciences, the National Academy of Engineering, and the Institute of Medicine. The members of the committee responsible for the report were chosen for their special competences and with regard for appropriate balance.

These proceedings were supported by the Electric Power Research Institute through contract number RP8011-21; the Department of Energy through grant number DE-FG02-93ER61703; the National Science Foundation through grant number IBN-9211798; the National Oceanic and Atmospheric Administration, U.S. Department of Commerce through grant number NA36RG0536; and the Presidents' Committee, National Research Council.

Any opinions, findings, conclusions, or recommendations expressed in this publication are those of the author(s) and do not necessarily reflect the views of the organizations or agencies that provided support for the project.

International Standard Book Number: 0-309-07188-7

Printed in the United States of America
Copyright 2000 by the National Academy of Sciences. All rights reserved.

THE NATIONAL ACADEMIES
National Academy of Sciences
National Academy of Engineering
Institute of Medicine
National Research Council

The **National Academy of Sciences** is a private, nonprofit, self-perpetuating society of distinguished scholars engaged in scientific and engineering research, dedicated to the furtherance of science and technology and to their use for the general welfare. Upon the a2uthority of the charter granted to it by the Congress in 1863, the Academy has a mandate that requires it to advise the federal government on scientific and technical matters. Dr. Bruce M. Alberts is president of the National Academy of Sciences.

The **National Academy of Engineering** was established in 1964, under the charter of the National Academy of Sciences, as a parallel organization of outstanding engineers. It is autonomous in its administration and in the selection of its members, sharing with the National Academy of Sciences the responsibility for advising the federal government. The National Academy of Engineering also sponsors engineering programs aimed at meeting national needs, encourages education and research, and recognizes the superior achievements of engineers. Dr. William A. Wulf is president of the National Academy of Engineering.

The **Institute of Medicine** was established in 1970 by the National Academy of Sciences to secure the services of eminent members of appropriate professions in the examination of policy matters pertaining to the health of the public. The Institute acts under the responsibility given to the National Academy of Sciences by its congressional charter to be an adviser to the federal government and, upon its own initiative, to identify issues of medical care, research, and education. Dr. Kenneth I. Shine is president of the Institute of Medicine.

The **National Research Council** was organized by the National Academy of Sciences in 1916 to associate the broad community of science and technology with the Academy's purposes of furthering knowledge and advising the federal government. Functioning in accordance with general policies determined by the Academy, the Council has become the principal operating agency of both the National Academy of Sciences and the National Academy of Engineering in providing services to the government, the public, and the scientific and engineering communities. The Council is administered jointly by both Academies and the Institute of Medicine. Dr. Bruce M. Alberts and Dr. William A. Wulf are chairman and vice chairman, respectively, of the National Research Council.

STEERING COMMITTEE ON OPPORTUNITIES FOR ADVANCEMENT OF ENVIRONMENTAL MARINE BIOTECHNOLOGY

David Manyak, CEO and President, Oceanix Biosciences, Hanover, MD
Judith McDowell, Woods Hole Oceanographic Institution, Woods Hole, MA
Roger C. Prince, Corporate Strategic Research Laboratory, Exxon/Mobil Research and Engineering Co., Annandale, NJ
Raymond A. Zilinskas, Monterey Institute of International Studies, Monterey, CA

Staff
Ralph B. Dell, Director, Board on Biology
Susan Roberts, Program Officer, Ocean Studies Board
Kathleen A. Beil, Administrative Assistant, Board on Biology
Susan Vaupel, Editor, Board on Biology
Marsha Williams, Project Assistant, Board on Biology

Preface

These proceedings summarize a workshop held October 5-6, 1999, to discuss the role of marine biotechnology in preventing degradation of the environment as well as in remediation and restoration. The agenda is reprinted in Appendix A. Each speaker summarized the current state of knowledge for each topic and highlighted the research needs in each area. Participants discussed the development of strategies for preventing or inhibiting biofilm development, remediation of oil spills and of marsh pollution, restoration of coral reefs, and the effects of heavy metals, overgrowth of microbes, and algal blooms. They also highlighted our critical knowledge gaps. Any advice, findings, conclusions, or recommendations are strictly those of the author and do not reflect a consensus of the workshop as a whole.

In addition to the speakers, those attending the workshop included representatives from the National Science Foundation, National Sea Grant Program of National Oceanic and Atmospheric Administration, Electric Power Research Institute, and Department of Energy, all of whom were sponsors of the workshop. We are indebted to these institutions for their sponsorship of the project and for their input during the workshop.

This report has been reviewed by persons chosen for their diverse perspectives and technical expertise in accordance with procedures approved by the National Research Council's Report Review Committee. The purposes of the independent review are to provide candid and critical comments that will assist the authors and the Research Council in making the published report as accurate as possible and to ensure that the proceedings accurately reflect the discussions at the workshop. The contents of the review comments and the draft manuscript remain confidential to protect the integrity of the deliberative process. We wish to thank the following persons for their participation in the review of this report: Keith Cooksey, Montana State University; Mark Hahn, Woods Hole Oceanographic Institution; Garrett Smith, University of South Carolina; and Lilly Young, Rutgers University Biotechnical Center for Agriculture and Environment.

Although the persons listed have provided many constructive comments and suggestions, responsibility for the final content of this proceeding rests solely with the speakers and discussants at the workshop.

Contents

Introduction and Goals 1
Roger C. Prince, Linda Kupfer, and Maryanna Henkert

Bacterial Biofilms and Biofouling:
Translational Research in Marine Biotechnology 3
Marc W. Mittelman

Antifouling 8
J. W. Costerton

Economic and Regulatory Aspects of Marine Biotechnology 14
Raymond A. Zilinskas

Policy Considerations for Advancing Marine Biotechnology 18
Lori Denno

Applications of Economics in the Field of Environmental
Marine Biotechnology 25
Diane Hite

Spilled Oil Bioremediation 34
Lily Young

In Situ Bioremediation of Oiled Shoreline Environments 44
Kenneth Lee

Contributions of Marine Biotechnology to
Marsh Oil Spill Restoration 61
Ralph J. Portier

Constraints on the Use of Bioremediation in Wetlands   68
Irving A. Mendelssohn

Restoration   73
Judith McDowell

Opportunities for Biotechnology for Coral and Reef Restoration   74
Aileen N. C. Morse

Coral Epidemiology   85
Laurie L. Richardson

Use of Trace Metals in Marine Bioremediation:
A Need for Fundamental Knowledge   96
François M. M. Morel

Microbial Contamination   102
Jed Fuhrman

Molecular Biology and Biotechnology in Marine Toxicology   112
Mark E. Hahn and John J. Stegeman

Critical Needs in Harmful Algal Bloom Research   126
JoAnn M. Burkholder

The Need for New Biotechnological Tools for
Conservation of Marine Environments   150
Michael Smolen

Social and Regulatory Aspects of the Marine Environment   154
Raymond A. Zilinskas

Rapporteur Comments on the Bioremediation Session   161
Roger C. Prince

Appendix A   171

Appendix B   175

Introduction and Goals

Roger C. Prince,[a] Linda Kupfer,[b] and Maryanna Henkert[c]

This 2-day workshop is the culmination of a study of the status and future of marine biotechnology. The overall goal of this workshop is to examine what was initially called "Opportunities for Marine Biotechnology in the United States," to consider where we are now in this field of "Environmental Marine Biotechnology," to envision the field in the future, and to discuss any impediments that might be encountered along the way. We hope that participants will address the question of where the federal government should invest its limited funds and what future initiatives should be planned.

The agencies that initially commissioned this study were the National Oceanic and Atmospheric Administration, National Sea Grant College Program, National Science Foundation, Department of Energy, and Electric Power Research Institute.

Marine biotechnology is coordinated at the federal level through the Office of Science and Technology of the President under the direction of the National Science and Technology Council, which has five Council

---

[a]Corporate Research Laboratory, Exxon/Mobil Research & Engineering Co., Annandale, NJ
[b]National Sea Grant College Program, OAR, National Oceanographic and Atmospheric Administration, US Department of Commerce, Silver Spring, MD
[c]Division of Molecular & Cellular Biosciences, National Science Foundation, Arlington, VA

committees that coordinate scientific activities throughout the federal government. The Committee on Science oversees marine biotechnology; the Subcommittee on Biotechnology and its Biotechnology Research Working Group oversee the Marine Biotechnology Task Force. This arrangement allows the Marine Biotechnology Task Force to meet as often as necessary, with a small group of interested people at the working level, to discuss current and planned activities, such as this workshop, and future initiatives to coordinate the federal investment in marine biotechnology.

The reports that follow highlight the most recent research results and leading edge ideas for applying the tools of biotechnology to the study of the marine environment. The discussions will surely help the responsible federal agencies to plan for future opportunities in research and applications of marine biotechnology.

Bacterial Biofilms and Biofouling: Translational Research in Marine Biotechnology

Marc W. Mittelman

Biological fouling ("biofouling") of engineered materials has been a significant problem for military and civilian oceangoing vessels. Materials deterioration, losses in heat-transfer efficiency, and mechanical blockages of fluid transport systems can result from biological fouling activities. These problems can also influence fuel consumption; for example, it has been estimated that 10% or more of fuel consumed by large naval vessels is required to overcome the viscous drag imposed by fouling organisms on ship hulls. In addition to the direct economic problems created by the activities of micro- and macrofouling organisms, the operational readiness of military vessels is influenced by the frequency of repairs and preventive maintenance activities that result from biological fouling.

The Office of Naval Research (ONR) has been the primary government-funding agency for biofouling research worldwide. Both basic and applied research have been supported under various ONR programs. Research to date has focused on the biology, ecology, detection, and treatment of putative fouling organisms. In addition, significant work has been funded in the fields of environmental toxicology and materials sciences.

Mechanisms associated with marine biofouling activities are, in most cases, identical to those seen in industrial fluid handling operations. Biological fouling is a major problem that results in significant environmen-

Altran Corporation, Boston, MA

tal impacts, both directly and indirectly through the misuse and misapplication of biocides. The biocide business in the United States is a multi-billion dollar business, and there are a number of Fortune 500 companies with products and services designed to control biofouling in industrial systems.

Bacterial biofilms (Figure 1) are the root cause of biofouling in most industrial systems. A biofilm is an agglomeration of bacteria on a surface that is surrounded or held together by extracellular polymeric substances. Bacteria produce extracellular polymeric substances, in part, to help them attach to surfaces and bind to one another. However, polymers also have a number of ancillary benefits, such as metal binding, which afford labile cellular components (e.g., sulfhydryl groups) some protection from otherwise toxic effects of heavy metals.

Due to their size and net negative charge, bacteria in solution act as colloidal particles. Their physicochemical behavior is much like that of clay particles, albeit clay particles with purposive behavior. A significant amount of ONR-sponsored research has focused on exploring the sort of intimate associations that exist between bacteria and various surfaces, particularly in marine environments. Understanding factors that promote the transition of bacteria (and other fouling organisms) from a planktonic to a sessile state is essential to the development of effective biofouling monitoring and treatment programs.

Biofouling involves the deleterious effects of microorganisms and some macroorganisms on engineered materials. These effects include

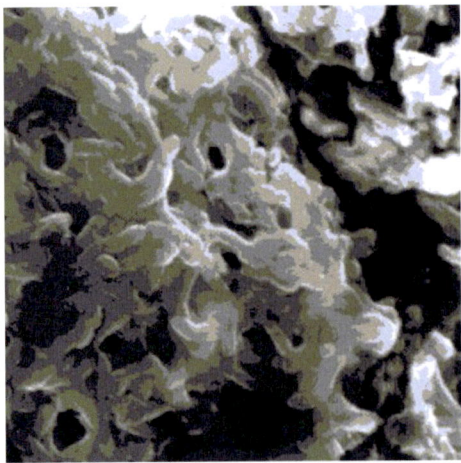

FIGURE 1. Bacterial biofilm on 316 stainless steel

mechanical blockages, significant losses in heat transfer, microbially influenced corrosion, product contamination, and threats to public health. Problems range from plugging of fire protection systems to microbially influenced corrosion of ferrous and nonferrous metals.

In addition to fouling problems in industrial systems, bacterial biofilms are responsible for significant problems in medicine, particularly with implanted medical devices. The limiting factor in the more widespread application of such critical devices is infection—rather than materials engineering considerations or surgical techniques. For example, devices such as the total artificial heart have a 90-day useful life span. The vast majority of patients chronically catheterized with indwelling devices develop urinary tract infections, usually within about 10 days after catheterization. Urinary catheter-related biofilms are the single greatest cause of nosocomial infections in hospitals, accounting for significant mortality and morbidity among hospitalized patients.

Marine research programs funded, for example, by ONR and the National Science Foundation have sponsored research into novel on-line detection mechanisms, physical treatments, development of antifouling compounds, adhesion-resistant surfaces, and antimicrobial coatings that might be useful for both industrial and medical applications. On-line detection techniques developed through research in the marine biotechnology arena have included evanescent wave technologies, fluorometry, acoustical monitoring, and electrochemical techniques. Evanescent wave technologies such as Fourier transform infrared and Raman spectroscopy have been applied to the detection of marine fouling organisms. A quartz crystal microbalance technology evolved from primarily Navy- and some Electric Power Research Institute-sponsored research as a way of evaluating on-line the development of organisms on surfaces. The US Navy has been very interested in developing on-line detection techniques that indicate both when the fouling problems are occurring and when to treat fouled surfaces.

The development of novel biofouling control measures had its origins in marine biotechnology research programs. Some of these programs have included antifouling treatments for ship hulls, pipelines, and marine structures. For example, the Navy has sponsored research into acoustical wave treatments involving high-frequency pressure transducers. A significant amount of work has also been devoted to so-called "natural products." Gorgonian coral is one source of animal-produced novel antifouling compounds; eelgrass is another example. Various extracts from marine animals and plants can be incorporated into antifouling paints and coatings, providing "natural" antifouling protection. These "natural product" antifoulants were discovered by marine biologists who observed that certain species of coral and marine plants were never colonized by bacteria, fungi, or higher organisms. In marine ecosystems, colonization space is a significant limiting factor for the development of many life forms. Therefore, the absence of colonizing marine organisms on these species of coral and marine plants was surprising. Further investigations revealed that gorgonian coral and eelgrass—among many others—produced complex organic compounds, which when extracted and applied to paints, similarly prevented colonization by fouling organisms.

A number of investigators have conducted research into fouling release compounds. Very often these are nontoxic compounds, such as silicones that are released from surfaces with exposure to fluid shear stresses. These types of ablative coatings are gradually sloughed under flowing conditions, taking with them attached fouling organisms. The association of fouling release compounds research with medical devices could be in the development of so-called biomimetic surfaces—mimicking natural tissue surface moieties. Surfaces exhibiting, for example, heparin-like moieties might retard microbial attachment and subsequent adhesion.

This brief discussion summarizes some of the key issues in this area of marine biotechnology research (Table 1). There is a wealth of information in the marine biotechnology arena, and the translation of much of this research into industrial and other environmental applications has yet to be realized.

TABLE 1. Key Issues in the Translation of Marine Biotechnology Research into Industrial, Environmental, and Medical Arenas

Issue	Challenge
Translation of marine research to industrial, medical, and environmental applications	Paucity of research into the microbial ecology of fouling biofilms
On-line monitoring for biofilms and biofouling	Sensitivity and selectivity of analytical tools
Novel antifouling compounds	Toxicity and materials compatibility
Commercialization of applicable marine biotechnology inventions	Intellectual property considerations; economics

REFERENCES

Bremer PJ, Geesey GG.
  1991   An evaluation of biofilm development utilizing non-destructive attenuated total reflectance Fourier transform infrared spectroscopy. Biofouling 3:89-100.

Characklis WG
  1990   Microbial biofouling control. In: Characklis WG, Marshall KC, eds. Biofilms. New York: John Wiley. p 585-633.

Costerton JW, Lewandowski Z, Caldwell DE, Korber DR, Lappin-Scott HM.
  1995   Microbial biofilms. Ann Rev Microbiol 49:711-745.

Davidson D, Beheshti B, Mittelman MW.
  1996   Effects of *Arthrobacter* sp., *Acidovorax delafieldii*, and *Bacillus megaterium* colonisation on copper solvency in a laboratory reactor. Biofouling 9:279-292.

Franklin MJ, Nivens DE, Vass AA, Mittelman MW, Jack RF, Dowling NJE, White DC.
  1990   Effect of chlorine and chlorine/bromine biocide treatments on the number and activity of biofilm bacteria and on carbon steel corrosion. Corrosion 47:128-134.

Gu JD, Roman M, Esselman T, Mitchell R.
  1998   The role of microbial biofilms in deterioration of space station candidate materials. Int Biodeter Biodegrad 41:25-33.

Hirota H, Okino T, Yoshimura E, Fusetani N.
  1998   Five new antifouling sesquiterpenes from two marine sponges of the genus *Axinyssa* and the nudibranch *Phyllidia pustulosa*. Tetrahedron 54:13971-13980.

Maki JS, Patel G, Mitchell R.
  1998   Experimental pathogenicity of *Aeromonas* spp. for the zebra mussel, *Dreissena polymorpha*. Curr Microbiol 36:19-23.

Marshall KC, Stout R, Mitchell R.
  1971   Mechanisms of the initial events in the sorption of marine bacteria to surfaces. J Gen Microbiol 68:337-348.

McCoy WF.
  1998   Imitating natural fouling control. Mat Perform 37:45-48.

Mittelman MW.
  1994   Emerging techniques for the evaluation of bacterial biofilm formation and metabolic activity in marine and freshwater environments. In: Morse D, ed. Recent Developments in the Control of Biodeterioration. London: Oxford University Press. p 49-56.

Mittelman MW.
  1997   Adhesion to biomaterials. In: Fletcher M, ed. The Molecular and Ecological Diversity of Bacterial Adhesion. New York: Wiley. p 89-127.

Mittelman MW.
  1997   Structure-function characteristics of bacterial biofilms in fluid processing operations. Guelph, Ontario: American Dairy Science Association.

Vrolijk NH, Targett NM, Baier RE, Meyer AE.
  1990   Surface characterization of two gorgonian coral species: Implications for a natural antifouling defense. Biofouling 2:39-54.

Antifouling

J. W. Costerton

Biofilms, which present problems in many different areas, result from the very tight adherence of bacteria to surfaces. Although bacteria are at risk and grow very poorly, when floating in water, they grow extremely well on surfaces. This characteristic has many consequences in different areas. I first studied the alpine environment, which probably has the cleanest water in any ecosystem, with sparse *Pseudomonas aeruginosa* growth of perhaps 8 bacteria/mL in the flowing water and abundant growth of at least 10^8 bacteria/cm^2 on the rocks. So it is in biofilms in any ecosystem that bacteria grow preferentially on surfaces.

A recent review in *Science* (Costerton and others 1999) illustrates the medical milieu to which Dr. Mittelman referred in his preceding introduction. The administration of antibiotics kills planktonic cells, which are also killed by specific antibodies and by white cells (Figure 1). We can deal with a certain number of planktonic bacterial cells in medical situations. If, however, biofilms have formed (which is the case much too regularly on foreign surfaces), then this biofilm is resistant to antibiotics and highly resistant to antibodies and white cells. The white cells are then unable to kill the bacteria because they cannot phagocytize them, and the "frustrated" white cells release enzymes that tend to digest the surrounding tissues. Thus a biofilm is essentially "bulletproof" against antibiotics and white cells. It withstands 1000 times as much antibiotic as floating (planktonic) cells of the same species. The Centers for Disease Control

---

Center for Biofilm Engineering, Montana State University, Bozeman, MT

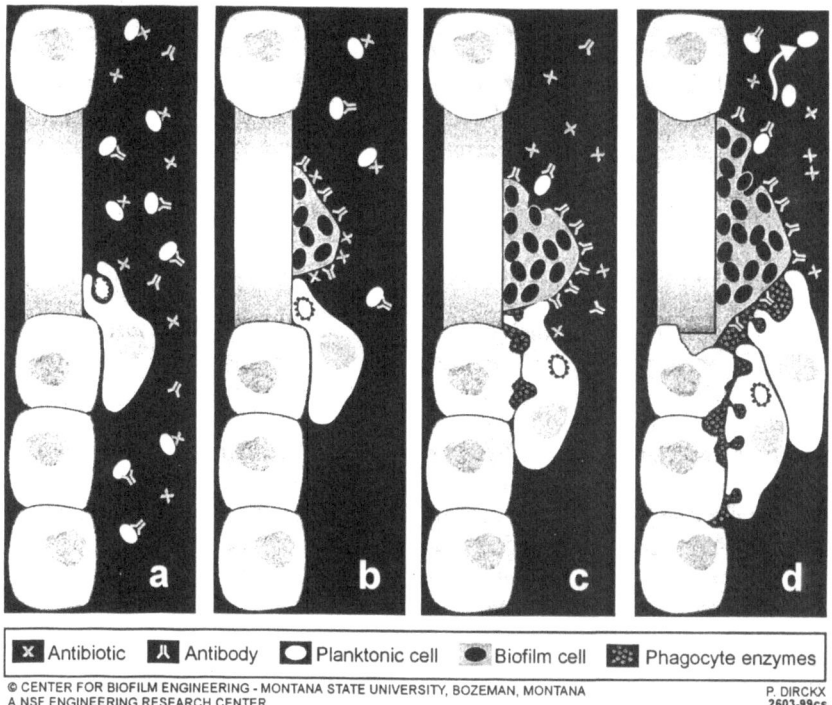

FIGURE 1. Diagrammatic representation of the very different effects of antibiotics and of host defense factors on bacteria growing in the planktonic mode of growth, as opposed to the biofilms mode of growth that characterizes very large numbers of modern bacterial infections. Reproduced with permission from *Science* (May 21, 1999).

have recently estimated that 65% of all infections seen by infectious disease specialists are biofilm infections (Costerton and others 1999).

Acute bacterial infections caused by planktonic bacteria have been largely controlled, primarily with antibiotics and vaccines; however, conditions such as children's middle ear infections persist until they are eventually outgrown. The persistent condition emerges as a very large concern. The NIH have recently issued three Requests for Applications (RFAs) in the biofilm area. In addition, and preceding the issuance of the RFAs, the Office of Naval Research and the engineering section of the National Science Foundation had funded research in this area.

Our ideas about biofilms have changed since 1993, when microbiologists viewed biofilms as a totally random accretion of cells on a surface.

In 1995, the idea of bacterial communication emerged and was called quorum sensing (Fuqua and others 1994). Bacteria floating in suspension were thought to produce signals. For example, in the case of *Pseudomonas aeruginosa*, two different signals are produced, and these simply diffuse through the cell envelope out into the surrounding milieu, and nothing happens. However, when planktonic cultures of bacteria reach a certain cell density (which Peter Greenberg calls a quorum), these signals begin to accumulate in relatively high concentration. They feed back through the membrane of the bacteria and activate their cognate receptor (R) protein to turn on the expression of certain genes. The first activation discovered was luminescence in *Vibrio* species, but the production of toxins, enzymes, and surfactants is also controlled by signals. Bacteria produce toxins, enzymes, and surfactant only when they achieve a certain concentration of cells. Some of the molecules that cause this controlled activity are called homoserine lactones (HSLs). Because we had a communication molecule and a phenomenon—biofilm formation—we then mutated the organisms so that they did not produce the communication molecule to learn what would happen to biofilm formation. The results were spectacular!

The wild type for a normal organism making both signals, the 3-oxydodecanoyl HSL signal and the butryl HSL, attaches to a surface, makes the exopolysaccharide, and eventually makes the biofilm. If we knock out the butyryl HSL, then it continues to make the polysaccharide and structured biofilms. If we knock out the 3-oxydodecanoyl HSL or both signals, the organisms that adhere to the surface cannot turn on polysaccharide production and cannot form biofilms. They adhere, but with a very small amount of surfactant or even with gentle stirring, they are removed from the surface. What was found, and published in *Science* in April 1998 (Davies and others 1998), was a signal that actually controls biofilm formation in *P. aeruginosa*.

People in the pharmaceutical industry are very interested in this area and in the possibility of developing resistance to biofilm blockage by HSLs. In the development of periodontitis, for example, the development of a deep trench between the tooth and the gum is caused not by one organism but by a whole community of microbes. People in the dental profession know that *Fusobacterium nacleatum* is the first microbe to get started; therefore, we are looking at the equivalent of the biofilm control HSL for this organism. The organism normally joins the plaque, which is perfectly harmless on the tooth but makes a deepening periodontal plaque by its presence. It is possible to simply reverse that process by adding the HSL-blocking analog to a mouthwash. *Fusobacterium* would thus be in the mouth, but it would not join the plaque deepening between the gum and the tooth. We thus would have a preventative measure for periodontitis.

An intriguing marine manifestation of signaling was discovered in Sydney, Australia, by an American (Peter Steinberg) and a Swede (Staffan Kjelleberg), and it became very successful commercially. The idea they followed was one that Drs. Manyak and Mittelman discussed earlier: that certain marine organisms do not allow bacterial biofilms to foul their surfaces. However, marine algae have the most acute problem because they cannot carry out photosynthesis if they develop biofilms on their surface. They become covered in slime, which attracts clay and buries them. There is a red algal species that grows all across Botany Bay in Sydney Harbor (called *Delcia pulcra*, meaning "delicious" and "beautiful") that occasionally fouls with macrophytes but virtually never with bacteria.

In the late 1980s, Staffan Kjelleberg and Pete Steinberg ground up some of these red algal fronds and found the active principle, a series of molecules called furanones. The furanones (which currently total 42) have been purified and added to boat coatings, to fishing nets (for tropical uses where they foul very badly), and to contact lenses. The latter use is the first medical application, which involves experimenting with resistance to bacterial fouling for long periods. This furanone panel (2) is located about 100 m from a sewage outfall in Sydney Harbor, where macrophyte fouling of control panels was evident 5 months after it was put in the base of the harbor, but the furanone-containing panel was virtually uncolonized. These natural furanone compounds have been shown (de Nys and others 1995) to block 3-oxydodecanoyl I ISL, specifically at the level of the interaction of the signal with its cognate receptor (R) protein. This discovery does not require synthesizing new compounds, but naturally occurring compounds can be used to affect biofilm formation.

I believe that furanones have actually been sold to a very large number of companies, and I think we will start seeing the emergence of medical devices containing furanones in the next 3 or 4 years. In addition, we will see the use of signals that trigger the detachment of planktonic bacteria from mature biofilms, as butyryl HSL triggers detachment in biofilms of *Pseudomonas aeruginosa*. Consider what happens then if we put the butyryl HSL in high-level concentrations into a biofilm produced by *P. aeruginosa*: We should be able to induce the detachment of those cells from that surface. This would be the anticipated result—a natural compound or an analog that triggered the mechanism and then locked the position covalently so that it stayed in a permanent detachment mode. We could take a formed biofilm of *Pseudomonas*, insert the detachment analog, and pull the whole biofilm off the surface to have floating cells

4 hours later. A clinical use of this technology could be found in the mechanical heart valve. The valve is attached to the endocardium by a sewing cuff made of a plastic fabric material. During infection, *Staphylococcus epidermidis* enters and makes a biofilm on the sewing cuff. We would administer low-level concentrations of the detachment signal, which would be a peptide, in the case of Gram-positive cells like *Staphylococcus epidermidis*. It would be possible to dissolve the biofilm even after an infection had taken place and after the sewing cusp had been colonized, and this maneuver would be done in the presence of very high-level concentrations of specific antibiotics.

Thus, I have described the two areas that we now understand in microbiology and biofilm microbiology—from planktonic cells coming onto a surface and forming aggregates and changing into the true biofilm with the signal-dependent production of matrix. We can experiment with all of those signals.

I would like to point out that the furanones must inhibit the biofilm formation signals for thousands of species of bacteria. This panel (Figure 2) is in a sewage-contaminated marine environment, and thousands of bacterial species are present in this contaminated environment, but none

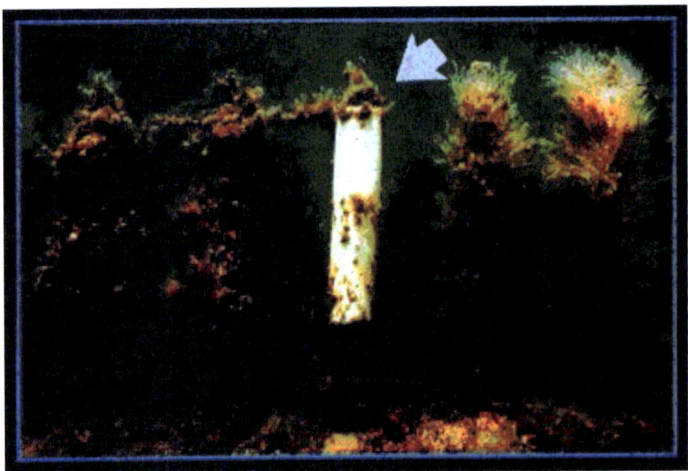

FIGURE 2. Photograph of a furanone-containing panel (center), with control panels, which had been exposed to the marine environment of Sydney Harbor for 5 months. Note the failure of marine biota in colonizing this material containing the natural blocking analog of the homoserine lactone signal that controls biofilm formation in many Gram-negative organisms. Reproduced with permission from Staffan Kjelleberg.

has become resistant and colonized the furanone panel. Also controlled by signals is the detachment event, in which these biofilm microcolonies simply dissolve and mobilize. We can now control these two activities, and a great deal of this work comes from the ONR program and from the NSF's concentration of resources in an engineering research center staffed by four-fifths engineers and one-fifth microbiologists in the Center for Biofilm Engineering (CBE). Together, these two groups are responsible for a very important discovery in medical, environmental, and industrial microbiology.

REFERENCES

Costerton JW, Stewart PS, Greenberg EP.
   1999   Bacterial biofilms: A common cause of persistent infections. Science 284:13218-13220.

Davies DG, Parsek MR, Pearson JP, Iglewski BH, Costerton JW, Greenberg EP.
   1998   The involvement of cell-to-cell signals in the development of a bacterial biofilm. Science 280:295-298.

de Nys R, Steinberg PD, Willemsen P, Dworjanyn SA, Gabelish CL, King RJ.
   1995   Broad spectrum effects of secondary metabolites from the red alga *Delisea pulchra* in antifouling assays. Biofouling 8:259-271.

Fuqua WC, Winans EP, Greenberg EP.
   1994   Quorum sensing in bacteria: The Lux R-Lux I family of cell density-responsive transcriptional regulators. J Bacteriol 176:269-275.

Economic and Regulatory Aspects of Marine Biotechnology

Raymond A. Zilinskas

INTRODUCTION

We all know that companies will engage in research and development that they believe will benefit themselves and their clients, so there is probably not a great need for governmental intervention in the industrial or commercial aspects of marine biotechnology. However, there are other activities that benefit society as a whole but are not likely to garner support from industry. Basic research is one such activity. Others are certain environmental applications that might serve to remediate degraded or polluted environments. Those who represent industry probably would not perceive that such activities would bring economic rewards in the short term, so there the government could have an important role. I hope that together we can identify areas where governmental intervention would be valuable.

With regard to regulatory aspects, the book titled *Genetically Engineered Marine Organisms: Environmental and Economic Risks and Benefits* (Zilinskas and Balint 1998) focuses on the risks and benefits that would attend the release of genetically engineered organisms in the marine environment, which is not something we are likely to address in detail in this workshop. Nevertheless, I believe there are certain lessons that can be drawn from what we who contributed to the book learned while researching and writing our chapters, especially with regard to the barriers that

---

Monterey Institute of International Studies, Monterey, CA

would hinder anyone who applies marine biotechnology to produce products that will be put into the seas.

One barrier pertains to the performance of an adequate risk assessment before something is introduced into the seas. When we began our study, our working hypothesis was that the risk assessment procedures used in the terrestrial environment would not be adequate for the marine environment. We eventually found that this was not the case. For example, the Environmental Protection Agency (EPA) has developed 21 points that have to be dealt with adequately by a developer of a genetically engineered organism before a risk assessment is performed (EPA 1990). We eventually concluded that these so-called "21 Points to Consider" were adequate and appropriate for the marine environment also. However, the big problem was that EPA's 21 points could not be satisfied if the proposed action involved the marine environment. In particular, the so-called "familiarity" criteria could not be met. In other words, due to a lack of data, no scientist can assert that he or she is sufficiently "familiar" with the marine environment or marine organisms to be able to assess risks inherent in releasing genetically engineered organisms into the open seas (Levin 1998). Clearly, basic research in marine biology and ecology has not yet generated the fundamental data needed to undertake a risk assessment of a proposed introduction of a genetically engineered organism into the marine environment. Because an adequate risk assessment cannot currently be performed, no release of a genetically engineered organism into the marine environment for any purpose is permissible in the US.

Another barrier is that a very difficult situation exists with regard to the regulatory regime dealing with marine environment. Most states have jurisdiction within 3 miles of the shoreline, although that distance is 9 miles for a few states. The federal jurisdiction is 3 to 200 miles. There is no area with joint authority.

However, the regulatory situation in the marine environment within the 200-mile limit is complex because there are potentially many different authorities that would govern any proposed activity in a given area. There are some real difficulties with, for example, ascertaining whether the EPA or the US Department of Agriculture would have regulatory authority if the proposed action involved introducing organisms into the open marine environment. In the terrestrial environment, the EPA usually has authority over microorganisms introduced deliberately on or over land for a purportedly beneficial purpose, such as environmental remediation, whereas the USDA governs introduced macroorganisms, such as genetically modified plants. Whether this situation will be reflected with regard to the marine environment is not clear because neither of these agencies has expertise in the marine environment, nor has Congress made known its view on the matter. So for now, federal agencies tend to ignore the prospect of marine biotechnology applications slated for the marine environment.

Surprisingly, we found the agency that apparently had the most jurisdiction in the marine environment is the US Food and Drug Administration (FDA). This is so because of the possibility that organisms and substances introduced into the seas might become food directly or could enter the food chain and thus become food eventually. FDA has gained some expertise in the marine environment because of their responsibility for seafood safety (Hazard Analysis and Critical Control Point regulations).

In addition to the state/federal problem, regional groups also enter into the regulatory picture. There are, for example, regional groups contributing to regulatory actions in the Chesapeake Bay Program and the Great Lakes' International Joint Commission. These groups would certainly express their opinion if someone proposed to introduce an organism into the seas under their jurisdiction, and possibly substances as well. It can be seen that the regulatory situation in the marine environment presents a kind of legal morass, capable of entrapping unwary developers who might wish to do something in the seas under national and international jurisdiction, as well as near and on shorelines.

It would appear that because so many regulations at all levels of government seek to address marine activities, there would be regulatory hurdles that would have to be overcome before any product of modern biotechnology, viable or inanimate, could be deliberately introduced into the marine environment (Stenquist 1998).

I hope that in this short presentation I have made clear that as we consider the types of biotechnological research pertaining to the marine environment that federal agencies ought to support, we also give some thought not only to possible applications of such research, but also to the regulatory barriers that might prevent applications from being realized. Why would an industry develop findings from basic research if it perceived that it would face great difficulty or uncertainty when actually trying to apply products or procedures? If this were the case, it probably would move on to develop applications for use in the terrestrial environment, where the regulatory situation is known. To overcome the regulatory barrier that might prevent applications in the marine environment, government agencies should consider sponsoring research in the social sciences that would clarify what the government could do to create a regulatory regime for the marine environment that is as unambiguous and certain as the existing regulatory regime in the terrestrial environment.

REFERENCES

EPA [US Environmental Protection Agency].
    1990   Points to Consider in the Preparation and Submission of TSCA Premanufacture Notices for Microorganisms. Washington, DC: US Environmental Protection Agency.

Levin M.
    1998   Risk assessment for uncontained applications of genetically engineered organisms. In: Zilinskas RA, Balint PJ, eds. Genetically Engineered Marine Organisms: Environmental and Economic Risks and Benefits. Amsterdam: Kluwer Academic Publishers. p 1-30.

Stenquist S.
    1998   Federal and state regulations relevant to uncontained applications of genetically engineered marine organisms. In: Zilinskas RA, Balint PJ, eds. Genetically Engineered Marine Organisms: Environmental and Economic Risks and Benefits. Amsterdam: Kluwer Academic Publishers. p 139-180.

Zilinskas RA, Balint PJ, eds.
    1998   Genetically Engineered Marine Organisms: Environmental and Economic Risks and Benefits. Amsterdam: Kluwer Academic Publishers.

Policy Considerations for Advancing Marine Biotechnology

Lori Denno

As the science of marine biotechnology advances, policy issues relating to resource access and management are emerging that may potentially affect the development of the field. With the support of the National Sea Grant College Program, the Center for the Study of Marine Policy at the University of Delaware has been involved in a 3-year study to research these issues. This research will be summarized in the Center's forthcoming book titled *Policy Issues in the Development of Marine Biotechnology* (Cicin-Sain and others 2000), which will contain components such as surveys of industry scientists and company representatives, evaluation of both the national and international policy frameworks that affect marine biotechnology, ways to structure relationships between industry and government to advance the field, and evaluation of public perceptions of the industry and biotechnology products.

Two basic questions trigger the policy context for marine biotechnology:

Where will the natural resources come from that will be used as models, studied, and developed? and

Where and how will these products be field-tested?

The first question relates to frameworks for marine resource access and management, and the second to emerging protocols for biosafety.

National Resources Conservation, Delaware Nature Society, Hockessin, DE

POLICY FRAMEWORK GOVERNING THE OCEANS AND MARINE RESOURCES

In the last few years, two important conventions have catalyzed the need for examination of relationships between marine resource management and the marine biotechnology industry: the Convention on Biological Diversity and the United Nations Convention on the Law of the Sea.

Convention on Biological Diversity

The Convention on Biological Diversity, often referred to as the Biodiversity Convention, was opened for signature in the course of the United Nations Conference on Environment and Development and entered into force on December 29, 1993, and remains without US Congressional ratification. The primary objectives of the Convention are "the conservation of biological diversity, the sustainable use of its components and the fair and equitable sharing of the benefits arising out of the utilization of genetic resources, including by appropriate access to genetic resources and by appropriate transfer of relevant technologies . . ." (UNCED 1993, p. 2). Measures to accomplish these objectives include the following: identifying and monitoring the components of biological diversity; providing for in situ biological diversity conservation through the establishment and maintenance of protected area systems; adopting economic measures that act as incentives for the conservation and sustainable use of biological diversity; increasing the emphasis on research and training and scientific and technical cooperation; and promoting public education and awareness regarding biological diversity (UNCED 1993, p. 2).

The need to link conservation and development of biodiversity as a key to ensuring incentives for conservation while expanding economic benefits is a primary focus of the Biodiversity Convention. It has broken new ground in international norms governing access to genetic resources by articulating the concept of regulating access to genetic resources to harness market incentives for the conservation of genetic information. In Article 15, Access to Genetic Resources, the Convention explicitly states that nations have sovereignty over their own genetic resources: "Recognizing the sovereign rights of States over their natural resources, the authority to determine access to genetic resources rests with the national governments and is subject to national legislation" (UNCED 1993, p. 8).

The Biodiversity Convention laid the groundwork for the establishment of national systems governing genetic resources. National rights may now be formally tied to genetic resources through regulations governing resource access.

United Nations Convention on the Law of the Sea

The Biodiversity Convention does not specifically address issues of access to marine genetic resources. However, the concept of sovereign rights over genetic resources in the marine environment is put forth in the United Nations Convention on the Law of the Sea (UNCLOS). Often referred to as a "constitution for the oceans," UNCLOS represents the culmination of 14 years of international negotiation to formulate and articulate rules to govern ocean space. Entered into force on November 16, 1994, UNCLOS delimits national ocean jurisdictions and sets forth rules governing the majority of ocean uses.

Part V of UNCLOS asserts that coastal nations party to the Convention may establish their own Exclusive Economic Zone (EEZ). The EEZ may extend up to 200 nautical miles from designated baselines. Within the EEZ, the coastal nation has sovereign rights over exploration, exploitation, management, and conservation of living and nonliving resources.

Conservation of living marine resources and protection of the marine environment are recurring themes throughout UNCLOS. This Convention supports the rights of nations to utilize and exploit marine resources within national waters; nonetheless, it also calls for "necessary measures" to ensure the ecological balance of the marine environment, both in areas under national jurisdiction and on the high seas (Vargas 1997).

Article 246, Section 3, stipulates that coastal States should grant consent for other States to carry out marine scientific research for the purpose of increasing scientific knowledge to benefit all mankind. Section 5, part A, however, specifically acknowledges that coastal States may withhold consent to the conduct of marine scientific research if the research "is of direct significance for the exploration and exploitation of natural resources . . ." (UNCLOS 1983, p. 87).

ACCESS TO MARINE RESOURCES IN NATIONAL WATERS

UNCLOS and the Biodiversity Convention together provide an international framework for access to marine genetic resources. Through these agreements, nations have sovereignty over marine resources out to 200 nautical miles, as well as the authority to determine conditions of access to biological and genetic resources that may ultimately possess commercial value.

Some companies have chosen to enter into benefit-sharing agreements with a host country or entity. In September 1991, Costa Rica's National Biodiversity Institute (INBio) announced an agreement with Merck & Co., Ltd., a US pharmaceutical firm, under which INBio agreed to provide Merck with chemical extracts from Costa Rica's conserved wildlands for

Merck's drug screening program in return for a two-year research and sampling budget and royalties on any resulting commercial products. INBio also agreed to contribute 10% of the budget and 50% of any royalties to the government's National Park Fund for the conservation of national parks in Costa Rica, and Merck agreed to provide technical assistance and training (Reid 1993). The agreement has been renewed twice for 2-year increments—in July 1994 and August 1996 (Vargas 1997).

In August 1997, Yellowstone National Park entered into a bioprospecting agreement with Diversa, Inc. Facilitated by the World Foundation for Environment and Development, it is the first agreement of its kind in the United States. Under the terms of the agreement, Diversa, a company specializing in discovery and application of enzymes, will conduct research to evaluate the bioactivity of thermophilic organisms found in Yellowstone. In return, Diversa will provide Yellowstone with an upfront payment of $100,000 over 5 years and a percentage of revenues generated by any products developed from research on samples taken from the Park (J. D. Varley, Yellowstone Center for Resources, personal communication, October 28, 1997).

To date, the use of such agreements in the marine environment has been limited, and marine resource collection is generally treated under UNCLOS provisions relating to marine scientific research. As new frameworks are developed, however, it is feasible that a coastal State could pass legislation stipulating conditions of access to its marine resources throughout the Exclusive Economic Zone. Highlighted as central to the Convention on Biological Diversity, such provisions might include technology transfer, recognition of indigenous knowledge, and benefit-sharing requirements.

ACCESS TO MARINE RESOURCES IN INTERNATIONAL WATERS

The Convention on Biological Diversity does not directly address biological resources in international waters but states that its Parties have a duty to cooperate in areas beyond national jurisdiction (UNCED 1993, p. 5). Currently, the primary governing mechanism in international waters is the International Seabed Authority, negotiated under the UNCLOS at a time when the mineral resources of the deep seabed were thought to have great economic potential. Due to their location, these resources were viewed as the "common heritage of mankind." The International Seabed Authority was thus created to ensure the equitable distribution of any benefits arising from mining operations. With the exception of certain fishing agreements, most other activities on the high seas are generally governed under the Freedom of the Seas regime, which supports unrestricted access to and use of ocean resources.

It is unclear which, if any, UNCLOS provisions may govern activities related to marine biotechnology. If the activity is viewed as "bioprospecting," it could be interpreted as a mining activity governed by the International Seabed Authority. If, however, it is interpreted as scientific research, as is typically the case to date, or as resource collection activity, such as fishing, then other policies will provide the access framework. This policy question is as yet unresolved, and it is worth noting that suggestions have been made for the creation of protected area networks on the high seas to protect the hydrothermal vent communities and other areas of high biological significance. Deep sea hydrothermal vents, home to many potentially industrially valuable microorganisms, exist not only in national waters, but also in areas where national jurisdiction is not clearly established, either because of conflicting claims or because the area is beyond national jurisdiction and falls within the purview of common heritage of mankind. Although access to resources in these areas is currently dependent on the availability of deep-sea technologies, it could, in the future, depending on the location of the vent area, be subject to Coastal State regulation by the Coastal State or by an international governing authority.

ISSUES OF BIOSAFETY

The Convention on Biological Diversity establishes the policy framework for biosafety issues. Addressed in the Convention, biosafety is the safe transfer, handling, and use of any living modified organisms (LMOs) resulting from biotechnology (UNCED 1993, p. 19.3). Moreover, the Convention calls for Parties to:

> "establish or maintain means to regulate, manage, or control the risks associated with the use and release of LMOs resulting from biotechnology which are likely to have adverse environmental impacts that could affect the conservation and sustainable use of biological diversity, taking also into account the risk to human health." (UNCED 1993, p. 11).

Key issues revolve around the lack of knowledge between how LMOs may interact with their environment, including competition with other species and their impact on nontargeted species in the ecosystem. Additionally, the perception exists that developing countries may be used as testing grounds for the release of LMOs.

The Biosafety Working Group was created under the auspices of the Convention in November 1995 to clarify and resolve these issues. Negotiations are still ongoing, but if and when a biosafety protocol is developed, it will likely contain provisions regarding transboundary movement of LMOs: advance informed agreement, risk assessment, and management; capacity building; information exchange; reporting and compliance; issues of liability and compensation; and socioeconomic considerations. Although enforcement of biosafety provisions may be challenging, such policies may be instrumental in cultivating public acceptance of biotechnology products and providing the clear regulatory frameworks for biotechnology firms that are necessary for sound investment strategies.

PUBLIC PERCEPTIONS OF BIOTECHNOLOGY

Another issue of relevance to the US marine biotechnology industry is public perception of the industry and its products. To date, much public interaction with biotechnology has taken place in a negative context. The most notable clashes between the public and the biotechnology industry have involved concerns voiced regarding field testing of genetically engineered microbial pesticides, levels of bovine growth hormone in dairy products, and genetically engineered agricultural products. It may be, however, that these concerns arose from perceived levels of risk resulting from a lack of scientific understanding or from inadequate communication between the industry and the public sector (Fleising 1991).

Benefits to society resulting from biotechnological processes are rarely as well publicized as the risks. For example, the public has virtually no way of differentiating a biotechnologically derived pharmaceutical product from one that has been manufactured through other methods. Yet, it is likely that few individuals would decline an important medical treatment based on its origin.

Public perception has the potential to affect the development of the industry. In a survey conducted by the Center for the Study of Marine Policy, 65% of industry representatives reported that ethical issues and related public perceptions could affect development of the field (Cicin-Sain and others, Forthcoming).

Opportunities for increased public awareness include outreach associated with benefit-sharing agreements, such as the Merck-InBio partnership. Such partnerships can serve as a vehicle to highlight important uses and applications of marine biotechnology and to demonstrate partnerships that meet both goals of economic development and conservation and sustainable resource use. Additionally, educational venues—schools and learning centers, such as science centers and aquariums—could provide platforms to teach the public about biotechnology and its contributions to sustainable resource use and the protection of human health.

In summary, both policy issues and social perspectives may affect the advancement of the marine biotechnology industry. Additional research is needed to help in further articulating some of the policy frameworks that govern access to marine resources and field testing of new products and processes and to provide additional insight regarding public response to and reception of new biotechnology products and processes.

REFERENCES

Cicin-Sain B. Knecht RW, Jang D, eds.
   2000   Policy Issues in the Development of Marine Biotechnology. Newark, DE: University of Delaware Center for the Study of Marine Policy. (Forthcoming).

Fleising U.
   1991   Public perceptions of biotechnology. In: Moses V, Cape RE, eds. Biotechnology: The Science and the Business 93. New York: Hardwood Academic Publishers.

Reid WV.
   1993.   Biodiversity Prospecting: Using Genetic Resources for Sustainable Development. Washington, DC: World Resources Institute.

UNCED [United Nations Conference on Environment and Development].
   1993   Convention on Biological Diversity. New York: United Nations Department of Public Information.

UNCLOS [United Nations Convention on the Law of the Sea].
   1983   The Law of the Sea with Index and Final Act of the Third United Nations Conference on the Law of the Sea. New York: United Nations Department of Public Information.

Vargas E.
   1997   Summary of Terms for the INBio-Merck & Col., Inc. Collaboration Agreement. <Evargas@quercus.inbio.ac.cr>.

Applications of Economics in the Field of Environmental Marine Biotechnology

Diane Hite

INTRODUCTION

The objective of this discussion is to review the current state of economics as it relates to the field of environmental marine biotechnology and to identify areas in which economics may play an important role. In general, physical scientists in many areas have not fully taken into account the economic implications of their research. This observation also applies to scientists involved in the study of marine biotechnology. Thus, there is almost no existing economics literature that relates to marine environmental biotechnology.

Most current economic research that deals with biotechnology in general has been primarily focused on topics that fall squarely in the realm of standard neoclassical economic analysis. Included in this area are topics such as patents and intellectual property rights (Bhat 1996), innovation (Audretsch and Stephan 1999; Mowery and Rosenberg 1998), the impact of biotechnology on industry structure (Acharya and Ziesemer 1996; Begemann 1997; Bijman 1996; Powell 1996), the ability of biotechnology to help increase food supplies (Rosegrant and Ringler 1997), and consumer acceptance of genetically modified foods (Caswell 1998). These fall in the areas of economics of innovation, industrial organization and agricultural, development, and consumer economics.

Department of Agricultural Economics, Mississippi State University, Mississippi State, MS

There is a need for a broader social economic approach to analyze the marine environmental biotechnology issues that are not currently being fully addressed. Social economics includes areas such as health economics, environmental economics, ecological economics, and public finance. A particularly useful analytical tool that has emerged from these fields is cost benefit analysis, which I argue will provide a useful tool to examine the economic issues inherent in marine environmental biotechnology.

BACKGROUND

Benefit cost analysis has become the standard method for determining the value of government projects and policies from a societal perspective. The objective is to establish all the potential costs and benefits that would be derived from a given project and then to determine whether the net outcome would have a positive impact. The benefits and costs should, as much as possible, reflect both tangible and intangible aspects of a project.

Neoclassical economics addresses the benefits and costs of a project that can be measured in existing markets for goods and services. For example, the benefit that individuals derive from consumption of fish can be measured by the demand for fish, and costs incurred by suppliers can be derived from the supply curve for fish. Thus, many aspects of a project can be directly measured by examining well-defined market behavior.

From a social economic point of view, a number of benefits and costs are derived from various projects and programs that are not valued in conventional economic markets. These are called externalities, or external costs and external benefits. Externalities are formally defined as unpriced outputs or inputs into a production process or other economic activity, and generally we use special techniques called nonmarket valuation to try to establish prices in such a case. The classic example of an external cost is air pollution, which is the byproduct of a smokestack industry.

The costs of pollution on society can range from morbidity and perhaps mortality resulting in lost worker productivity, to the lost sense of well-being individuals may feel from reduced visibility. Air pollution externalities of this type have been studied by a number of environmental economists (Brookshire and others 1982; Freeman 1974; Schulze and others 1998).

If all of the externalities and spillovers from a project are not accounted for, projects that are potentially worthy of funding may not be deemed economically feasible. However, many projects are undertaken in which associated externalities are not accounted for, resulting in devastating negative economic impacts. Misguided agricultural policies in less developed countries illustrate this point well—for instance, World Bank funding of US style farm techniques in Africa has been recognized as an exacerbating factor in desertification and microclimate change.

APPLICATION OF BENEFIT COST ANALYSIS TO BIOTECHNOLOGY IN MARICULTURE

There are a number of applications for benefit cost analysis in environmental marine biotechnology. One example is the analysis of mariculture of genetically modified marine species (Hite and Gutrich 1998). I present some of the basic concept of that paper here. Central to the analysis is the idea that a number of the issues addressed here should be viewed as increments to costs and benefits that already exist in the marine aquaculture industry. It should be noted that genetic modification is a relatively minor player in biotechnology, and the following analysis is meant only to illustrate the significance of the economic concepts of spillovers and externalities.

Neoclassical Economic Benefits and Costs

From a neoclassical economic standpoint, potential benefits would include increased growth rates of maricultured macroorganisms. Because some modified species can mature in 67% of the normal time, reach sizes up to 11 times that of their natural counterparts, or both, an increase in food supply would result. The effect would be enhanced by potential improvement to marine plant and animal health.

The primary costs that accrue to such an enterprise are related to regulatory and containment costs to prevent accidental releases. Included in containment costs are costs of increasing the strength of cages and securing facilities beyond that already experienced in conventional mariculture. Forster (1996) reports the cost of aquaculture cages ranges from $10 to $100/m^3$, with the most expensive cages providing the most containment protection, but suggests that aquaculture would become unprofitable with cage prices above $50/m^3$. The cost of monitoring to avoid accidental releases would potentially be extremely high and could be particularly difficult to implement; in 1995, expenditures for enforcing all environmental regulations amounted to approximately $115 billion.

Costs from a Social Economic Perspective

External or social costs associated with an activity can be significant, as are some of those associated with current mariculture practices. For instance, large scale farming of marine species may damage the benthic layer of the ocean. Thus, to the extent that the introduction of genetic modifications would encourage the growth of the mariculture industry, an additional external cost would be incurred.

A second possible external cost that might accrue is the contribution of wide-scale mariculture in exacerbating problems associated with resistant microorganisms. Microbe resistance presents an enormous risk for human health and creates problems for the mariculture industry in that diseases of macroorganisms held in close quarters may create a significant financial risk.

A third consideration is the concept of pecuniary externalities that arise from increased supplies of fish on the market as a result of genetically enhanced mariculture. Regulatory distortions in the worldwide fishing industry have arisen as a result of attempts to curtail overfishing by limiting fishing seasons. Fleets have reacted by investing heavily in expensive capital equipment, resulting in increases in fleet size from about 2.2 million to 3.3 million vessels worldwide from 1970 to 1989 (Powers 1995), and tonnage has nearly doubled (FAO 1995). The end result is that if increased supplies of fish result from genetic modification in mariculture, prices should fall. In response to lower prices, the existing fishing industry would have a perverse incentive to apply more effort in natural fisheries, depleting natural stocks at a rate faster than currently experienced.

Finally, there are external costs associated with the risk of accidental release of a genetically modified species. To assess such a cost, it is necessary to take into consideration what would become of an organism should it escape, that is, to determine whether it will be able to survive out of captivity and if so, fill some ecological niche and disrupt the biodiversity of the marine environment and become established as an exotic species. Examples of exotic marine species disrupting ecosystems and economies are abundant (e.g., release of an exotic ctenophore in the Black and Azov seas that led to the collapse of local fisheries [Travis 1993]). However, the best-known case is perhaps that of *Dreissena* (zebra mussels). It has been estimated that as of the year 2000, this species will have incurred economic costs of $3 billion to $5 billion annually (ASNTF 1992; Cohen and Carleton 1995). In addition, this species has the ability to drastically alter local freshwater habitats by changing water clarity. Habitat disruption may have even more serious economic implications, as discussed below.

Benefits from a Social Economic Perspective

Significant beneficial externalities would accrue to genetic enhancement of food fish. Included among these benefits are the potential for reduced pressure on natural fisheries and perhaps preserving some of the biodiversity of the marine environment. This latter benefit is certainly contingent on the future regulation and reaction of the world's fishing fleets, as noted previously. This is a particularly significant benefit because seafood demand is expected to increase 70% in next 35 years (JSA 1992; NSTC 1995), and in six of 11 fishing regions, more than 60% of species have already been depleted or fished to their biological limit (FAO 1995). This fact has additional economic implications for labor markets; for instance, 40,000 jobs were lost in eastern Canada in 1992 as a result of the collapse of the Atlantic cod fishery (Clayton 1995; Garcia and Newton 1994; WRI 1996).

BROADER APPLICATIONS OF ECONOMICS IN ENVIRONMENTAL MARINE BIOTECHNOLOGY

During the National Research Council Workshop on Environmental Marine Biotechnology, a number of areas of interest for research in various areas of physical and marine sciences were broached. In this section, I discuss the role of economics in justifying and enhancing these areas of research. I address the three broad interest areas discussed in this workshop—biomaterials, bioremediation, and restoration.

Biomaterials

In the area of biomaterials, it appears that the development of materials dealing with biofilms will have a very significant benefit from a societal standpoint. In particular, Dr. Costerton emphasized that 65% of all infections are biofilm infections that are related to the formation of polysaccharides and that such infections are highly resistant to antimicrobials. Dr. Costerton suggested ways in which development of polysaccharides could be blocked, thereby preventing a number of serious infections. From a health economics perspective, this research could have major repercussions in that it can provide new means to control human bacterial infections without the development of new antibiotics. The current state of drug-resistant microbes has been very costly to society—the cost of antibiotic resistance has recently been estimated at $30 billion per annum in the United States (Spake 1999). It is unclear what is included in this cost estimate, but it would be increased substantially by costs attributable to lost research and development investment. It is probable that the time spent on developing new antibiotics is longer than the time it takes for them to lose effectiveness. In addition, the ineffectiveness of antibiotics contributes to lost economic productivity from labor force morbidity and mortality, and can help to escalate medical costs.

Dr. Mittelman brought up another aspect of the biofilms-health interface when he emphasized the role of biofilms in harboring bacteria that are potentially harmful to health (e.g., *Legionella*). Once again, from Dr. Mittelman's perspective, biofilms research would have major applications in prevention of microbial infection, particularly in instances where surface adhesion of biofilms poses infection risk, such as in artificial hearts and urinary tract infections associated with catheterization. Once again, the spillover benefits of controlling biofilm infections would be great and would present possibilities for prevention of nosocomial infections that contribute directly to hospital costs and indirectly to economic productivity. Dr. Manyak also touched on the health care theme when he discussed the use of marine proteins in medical device applications.

Bioremediation

Bioremediation is one area in which the benefits of research are quite obvious. Bioremediation can represent huge cost savings to firms that are responsible for cleaning up oil spills, for instance. Bioremediation may also increase the recovery rate of valuable fisheries after a spill, providing economic benefits to local economies. In addition, bioremediation may provide a means to clean oil spills more thoroughly, leading to restoration of a larger variety of species than might be otherwise unattainable.

Dr. Young presented research on ways to biodegrade petroleum in estuarine sediments. Deposition of sediments has limited access to many ports, and toxic substances in the sediments have rendered other types of remediation, such as dredging, impractical. By finding ways to eliminate toxics from sediment, dredging could once again be used to open ports to a wider variety of ship traffic. The economic benefits could be measured directly, in terms of job creation and other economic activity. However, from an economic justice standpoint, such research could have positive repercussions if inaccessible ports have contributed to economic decline in an inner city such as in Newark, New Jersey (Economics faculty, NJ Institute of Technology, 1996, personal communication).

Other aspects of marine bioremediation as discussed by Drs. Lee, Porter, and Mendelssohn have important economic components. The most important area to be considered is the contribution of bioremediation to overall ecosystem health. In the case of residual hydrocarbons, a relevant question is: How will the health of a fishery be impacted? The same question applies in the case of marsh remediation; marshes are biologically sensitive areas that house nurseries and spawning grounds for a number of marine species. Thus, economic spillovers would accrue to areas offsite from the marsh area. The significant economic impact of coastal wetlands remediation and cleanup on recreation and coastal real estate values, mentioned earlier in this session, is external to the cleanup itself. The value of worldwide coastal tourism alone increased 20-fold from 1950 to 1995—to $60 billion per year—and is expected to double again by 2010 (McGinn 1999).

Restoration

Drs. Morse and Richardson discussed the restoration of coral reefs and the diseases and other factors that are affecting reef health and viability. Reefs are known to provide a large number of ecological services to the marine environment, and loss of reefs poses a potentially serious threat to that environment. One example of an economic cost of reef loss is in the subsequent loss of other marine species that rely on the reef during some or all of their life cycles. Another is their significant role in the livelihood of food fish species.

Direct economic benefits of reefs may be measured by local tourism expenditures. Bonaire Marine Park, for example, generates annual gross revenues of $23.2 million/annum for dive-based tourism (Dixon and others 1993, 1994; Scura and Van't Hof 1993). Many of the world's reefs are in less developed countries or in small countries that base a large portion of their national economy on tourism based on reefs.

REFERENCES

Acharya R, Ziesemer T.
    1996   A closed economy model of horizontal and vertical product differentiation: The case of innovation in biotechnology. Econ Innovat New Technol 4:245-264.
ASNTF [Aquatic Species Nuisance Task Force].
    1992   Proposed Aquatic Nuisance Species Program. Alexandria, VA: US Fish and Wildlife Service.
Audretsch DB, Stephan PE.
    1999   Knowledge spillovers in biotechnology: Sources and incentives. J Evolut Econ 9:97-107.
Begemann BD.
    1997   Competitive strategies of biotechnology firms: Implication for US agriculture. Agric Appl Econ 29:117-22.
Bhat MG.
    1996   Trade-related intellectual property rights to biological resources: Socioeconomic implications for developing countries. Ecolog Econ 19:205-217.
Bijman JW.
    1996   Biotechnology and vertical integration in the Dutch potato chain. In: Galizzi G, Venturini L, eds. Economics of Innovation: The Case of Food Industry. Contributions to Economics. Heidelberg: Physica.
Brookshire DS, Thayer M, Schulze WD, D'Arge RC.
    1982   Valuing public goods: A comparison of survey and hedonic approaches. Am Econ Rev 72:165-178.
Caswell JA.
    1998   How labeling of safety and process attributes affects markets for food. Agric Resource Econ Rev 27:151-158.

Clayton M.
  1995   A Fish Tale? Canada Tries to Save Stocks While Overfishing. Christian Science Monitor, March 20, 1995. p 52.
Cohen AN, Carlton JT.
  1995   Non-indigenous aquatic species in a United States estuary: A case study of the biological invasions of the San Francisco Bay and delta. Report to the US Fish and Wildlife Service. Washington, DC: GPO.
Dixon JA, Scura LF, Carpenter RA, Sherman PB.
  1994   Economic Analysis of Environmental Impacts. London: Earthscan Publications, Ltd.
Dixon JA, Scura LF, Van't Hof T.
  1993   Meeting ecological and economic goals: Marine parks in the Caribbean. Am Bio 23:117-125.
FAO [Food and Agriculture Organization of the United Nations].
  1995   Review of the state of world fishery resources: Marine fisheries. FAO Circular No. 884. Rome: FAO.
Forster J.
  1996   Cost and marker realities in open water aquaculture. Proceedings of the Open Ocean Aquaculture Conference cosponsored by the University of Maine and the University of New Hampshire, Portland, OR, May 8-10, 1996.
Freeman AM, III.
  1974   On estimating air pollution control benefits from land value studies. J Environ Econ Managemt 1:74-83.
Garcia S, Newton C.
  1994   Current situations, trends, and prospects in world capture fisheries. Paper presented at the Conference on Fisheries Management: Global Aspects, Seattle, WA.
Hite D, Gutrich J.
  1998   An economic analysis of introduced marine genetically engineered organisms. In: Zilinskas RB, Balint PJ, eds. Marine Biotechnology: Assessing and Managing the Economic and Ecological Risks. The Netherlands: Kluwer.
JSA [Joint Subcommittee on Aquaculture].
  1992   Aquaculture in the United States: Status, opportunities and recommendations. A report to the Federal Coordinating Council on Science, Engineering, and Technology. Washington, DC: USDA.
McGinn AP.
  1999   Safeguarding the Health of the Oceans. Worldwatch Paper 145. Washington, DC: Worldwatch Institute.
Mowery DC, Rosenberg N.
  1998   Paths of Innovation: Technological Change in Twentieth-Century America. Cambridge: New Cambridge University Press.
NSTC [National Science and Technology Council; Office of Science and Technology Policy].
  1995   Biotechnology for the 21st century: New horizons. Washington, DC: GPO.
Powell WW.
  1996   Inter-organizational collaboration in the biotechnology industry. J Inst Theoret Econ 152:197-215.
Powers D.
  1995   New frontiers in marine biotechnology: Opportunities for the 21st century. In: Lundin CG, Zilinskas RA, eds. Marine Biotechnology in the Asian Pacific Region. Stockholm: World Bank.

Rosegrant MW, Ringler C.
  1997 World food markets into the 21st century: Environmental and resource constraints and policies. Aust J Agric Resource Econ 41:401-428.
Schulze WD, McClelland GH, Lazo JK, Rowe RD.
  1998 Embedding and calibration in measuring non-use values. Resource Energy Econ 20:163-178.
Scura LF, van't Hof T.
  1993 Economic feasibility and ecological sustainability of the Bonaire Marine Park. Environment Department Positional Working Paper 1993-44. Washington, DC: World Bank.
Spake A.
  1999 Losing the Battle of the Bugs. US News & World Report, May 10, 1999. p 52.
Travis J.
  1993 Invader threatens Black Azov seas. Science 262:1366-1367.
WRI [World Resources Institute].
  1996 World resources guide to the global environment. New York: Oxford University Press.

Spilled Oil Bioremediation

Lily Young

Although there are few catastrophic oil spills, there are many contaminated harbors. The sediment in these contaminated harbors typically has a very small aerobic layer where you can see the oxidized iron; however, there is no oxygen under those few centimeters in this sticky, mucky, smelly anaerobic anoxic sediment. This means that if the levels of polynuclear aromatic hydrocarbons (PAHs) are not dispersed or degraded in the water column or aerobic zone, they accumulate in this very large anoxic sediment reservoir.

Table 1 is from a publication by Huntley and others (1995), which summarizes the PAHs accumulated in the sediment of various sites in and around the New York-New Jersey harbor. Clearly, the PAHs are still in the sediment and have not disappeared. They may still be there because of limiting nutrients or, more probably, a lack of oxygen. The question is whether there is a biotechnology fix for this—perhaps.

When we look at some of the crude and refined petroleum oils and the constituents, alkanes, cycloalkanes, and aromatics, we see that all are biodegradable by a variety of microorganisms, by certain groups of mostly aerobic organisms. They use oxygen in their metabolism, which is important. For instance, there is an aerobic pathway for naphthalene that is metabolized to salicylates (salicylic acid), and this is further broken down

---

Biotechnology Center for Agriculture and Environment, Rutgers University, New Brunswick, NJ

TABLE 1. Selected Polynuclear Aromatic Hydrocarbons (PAHs) and Petroleum Hydrocarbons in NY/NJ Sediment (mg/kg dry wt ± s.d.)[a]

Chemical	Arthur Kill	Newark Bay	Passaic River
Acenaphthene	2.4 (6.88)	2.3 (5.45)	13 (93)
Anthracene	2.3 (6.12)	2.4 (5.57)	8.3 (47)
Benz [a]anthracene	1.6 (2.46)	2.2 (4.74)	7.3 (30.5)
Benzo[a]pyrene	1.4 (2.19)	1.7 (4.22)	5.5 (20.1)
Benzo[k]fluoranthene	1.3 (1.84)	1.7 (4.20)	3.9 (11.2)
Chrysene	1.9 (2.79)	2.2 (4.90)	7.9 (32.2)
Dibenz [a,h]anthracene	0.68 (0.27)	1.5 (4.14)	2.2 (5.21)
Dibenzofuran	0.88 (1.12)	2.0 (4.60)	2.6 (8.04)
Fluoranthene	5.2 (13.2)	5.3 (15.0)	10 (40.2)
Fluorene	1.4 (3.34)	2.1 (5.60)	7.7 (49.3)
2-Methylnaphthalene	1.6 (4.36)	1.5 (4.15)	11 (80)
Naphthalene	3.0 (10.9)	1.8 (4.44)	16 (123)
Phenanthrene	5.8 (18.0)	4.7 (15.8)	19 (113)
Pyrene	3.6 (5.59)	3.9 (9.76)	14 (61.9)
TPAH	37 (81.7)	44 (98.5)	145 (739)
TEPH	703 (1184)	339 (535)	1,520 (5,970)

[a]Adapted from Huntley SL, Bonnevie NL, Wenning RJ. 1995. Polycyclic aromatic hydrocarbons and petroleum hydrocarbon contamination in sediment from the Newark Bay estuary. Arch Environ Contam Toxicol 28:93-107.

to constituents that are then incorporated into central metabolic pathways (Sutherland and others 1995; Figure 1).

An important point is that the key enzymes that activate this very stable bicyclic molecule—the dioxygenases—require molecular oxygen as a reactant. Oxygen therefore must participate in the reaction to activate the rings or to catalyze the ring fission that occurs here. Salicylic acid requires an oxygenase to break the ring. Hence, oxygen is a reactant.

In the typical known pathway for alkane degradation, which produces fatty acids that can then be incorporated into basic metabolic processes, monooxygenases are required. Again, oxygen is one of the reactants in the reaction that forms fatty acids.

The preceding description sets the stage. In the aerobic environment, we have organisms that require the activity of oxygenases; and in an anoxic environment, where oxygen is not present, let us consider what happens. If anything happens, it must occur through a significantly different biochemical and metabolic mechanism. That area is where I have involved my students, showing them that we can find these organisms because we know they are there. We know they can carry out certain novel degradation reactions, but we do not yet know whether this process is relevant to the environment.

The anaerobic organisms found in that large reservoir of anoxic sediment do not use oxygen but can use other inorganic electronic acceptors

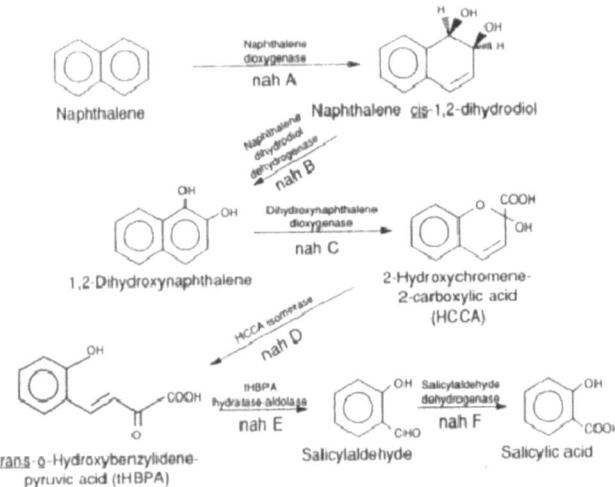

FIGURE 1. Initial steps in the metabolism of naphthalene to salicylic acid by *Pseudomonas putida*. The genes coding for enzymes involved in the metabolism of naphthalene are designated by nah. Reprinted with permission from Sutherland JB, Rafii F, Kaha AA, Cerniglia C. 1995. Mechanisms of polycyclic aromatic hydrocarbon degradation. In: Young LY, Cerniglia CE, eds. Microbial Transformation and Degradation of Toxic Organic Chemicals. New York: Wiley-Liss. p 269-306.

for respiration. Aerobic organisms use oxygen, and anaerobic organisms, as Dr. Costerton mentioned, can use nitrate. Certain types of microbes also use sulfate and carbonates. This ability is specific. Some organisms, like the denitrifiers, can also use nitrates if oxygen is not available. Others such as the sulfate reducers can only use sulfate and are strict anaerobes. Those that can use carbonate to form methane are called methanogens and are strict anaerobes. We know that these organisms are very important in the general carbon cycle. However, we know much less about whether these groups of organisms have a role in terms of contaminant degradation in these anoxic environments.

It may be helpful to review some of the work we have done. In one case, we looked to see whether contaminated sediment from the New York-New Jersey harbor contained anaerobic organisms that can degrade any of these polycyclic aromatic compounds. We first looked at naphthalene, 2-methylnaphthalene, phenanthrene, and pyrene, and then at some oxidized derivatives such as 1-naphthol and 1-naphthalene. Everything was handled anaerobically; that is, there was no oxygen available. So the only electron acceptor available was either nitrate, Fe(III), sulfate, or carbonate. If any organisms could use these compounds (PAH) as carbon and could use these electron acceptors for respiration, then we would eventually see activity.

Among the real PAHs (not the oxidized derivatives)—naphthalene, methylnaphthalene, phenanthrene, and pyrene—the first three compounds could be metabolized using sulfate to support the metabolism. It took months for the activity to be observable—3 months were usually adequate, but 9 months were necessary in some cases. Nonetheless, we succeeded in selecting for anaerobic organisms that use PAHs in the absence of oxygen.

The time courses for naphthalene and phenanthrene can be seen in Figure 2 (Zhang and Young 1997). After the initial lag period of several

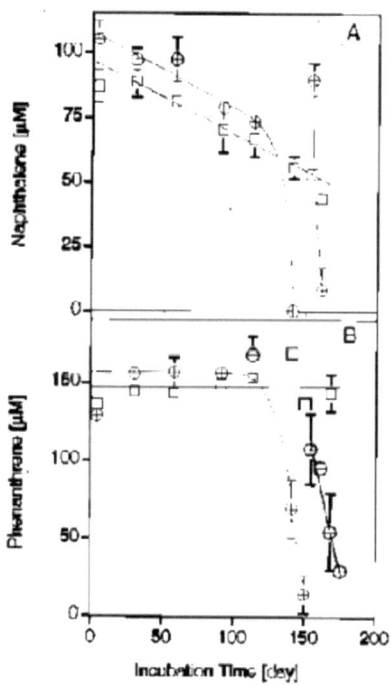

FIGURE 2. Initial degradation of NAP (A) and PHE (B) in 10% AK sediment-inoculated, sulfate-reducing enrichments. The slow decline of the NAP concentration in the autoclave controls is due to volatile loss during sampling. Data points represent the means of three replicates for active cultures (⊗) and the means of two replicates for autoclaved controls (□). Reprinted with permission from Zhang X, Young LY. 1997. Carboxylation as an initial reaction in the anaerobic metabolism of naphthalene and penanthrene by sulfidogenic consortia. Appl Environ Microbiol 63:4759-4764.

months, activity was relatively rapid. One of the ways we can determine whether true metabolism has taken place is to use radiolabeled compounds (Table 2) and determine how much of the carbon in the radiolabeled naphthalene and phenanthrene can be recovered as CO_2 (Zhang and Young 1997). Our results indicate that the radiolabeled substrates can be converted to CO_2 to the extent of 89 and 92%. Hence, most of the substrate is being degraded to CO_2.

We used stable isotope C^{13}-labeled compounds and deuterated compounds to prove to ourselves that the metabolites we were seeing actually came from the substrates we gave them. Up to this point, we did not know what was happening between what we started with and when we ended up with CO_2. We can now partially answer what happens in between with the following observations. In summary, we are able to show that 2-methylnaphthalene and naphthalene can both be converted to 2-naphthoic acid during degradation to carbon dioxide (Zhang and Young 1997). Did this metabolite actually come from the parent substrate or did it come from somewhere else? We can answer this question by using deuterated methylnaphthalene, and we can show by gas chromatography-mass spectrometry that deuterated naphthoic acid is produced. This evidence indicates that the microorganisms were carrying out the degradation of naphthalene through a mechanism that is very different from aerobic organisms.

We have also been able to show that the carboxylation of the naphthalene occurs through an inorganic carbonate addition to the molecule, using a stable isotope C^{13}-carbonate in solution. After incubation and gas chromatography-mass spectrometry analyses, we showed this carboxyla-

TABLE 2. Mineralization of [^{14}C]Polynuclear Aromatic Hydrocarbons ([^{14}C]PAHs) in Acclimated, Sulfidogenic Consortia[a,b]

PAH Tested	Total added radioactivity (dpm)[c]	Amount of radioactivity (dpm) Recovered as $^{14}CO_2$	Left in slurry	Total radioactivity recovered (%)
NAP	107,170	95,671 (89.3%)	4,350 (4.1%)	93.3
PHE	529,076	487,057 (92.1%)	27,709 (5.3%)	97.3

[a]Reprinted with permission from Zhang X, Young LY. 1997. Carboxylation as an initial reaction in the anaerobic metabolism of naphthalene and penanthrene by sulfidogenic consortia. Appl Environ Microbiol 63:4759-4764.
[b]For each PAH, six replicate samples were established and 150 µM unlabeled PAH was added. Radiolabeled PAH was then added to three of the six replicates. Samples without radioactive PAH were analyzed to monitor the progress of the PAH degradation. Incubation lasted 24 and 42 days for NAP and PHE, respectively, at room temperature (24 ± 2°C).
[c]Standard deviations of the reported numbers were within a 2% range ($n = 3$).

tion occurring. The standard mass spectrum of 2-naphthoic acid has major fragments at 244, 229 and 185, 155. After using C^{13}-carbonate, the major fragments all increase by one mass unit, namely, 245, 230, 156 (Zhang and Young 1997; Figure 3).

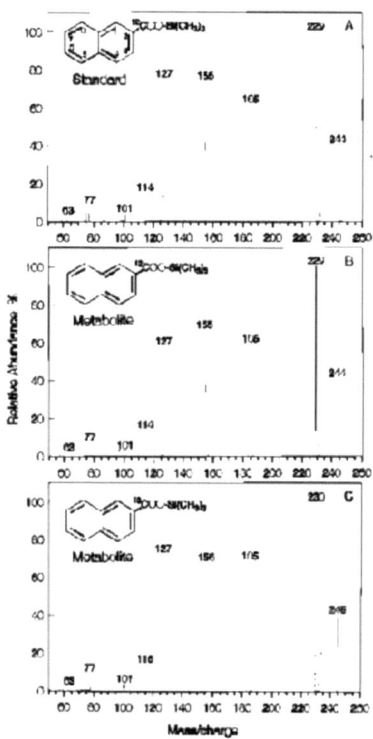

FIGURE 3. Mass spectra of trimethylsilyl derivatives of a 2-NA standard (A) and 2-NA extracted from the sample supplemented with NAP and either [^{12}C]bicarbonate (B) or [^{13}C]bicarbonate (C). The mass spectrum of the 2-NA standard (A) contains five major peaks: m/e 244, 229, 185, 155, and 127. The m/e 244 peak represents the molecular ion. The fragmentation ion of m/e 229 is the result of the loss of a –CH$_3$ group (244 – 15 = 229); the fragmentation ion of m/e 185 is from the loss of a –CH$_3$ group and a –COO group (244 – 59 = 185); the fragmentation ion of m/e 155 is from the loss of an –OSi(CH$_3$)$_3$ group (244 – 89 = 155); and the fragmentation ion of m/e 127 is from the loss of a –COOSi(CH$_3$)$_3$ group (244 – 117 = 127). The identification of the 2-NA metabolites in panels B and C is based on comparison of the GC retention time of the derivatized standard for 2-NA (10.41 min). Reprinted with permission from Zhang X, Young LY. 1997. Carboxylation as an initial reaction in the anaerobic metabolism of naphthalene and penanthrene by sulfidogenic consortia. Appl Environ Microbiol 63:4759-4764.

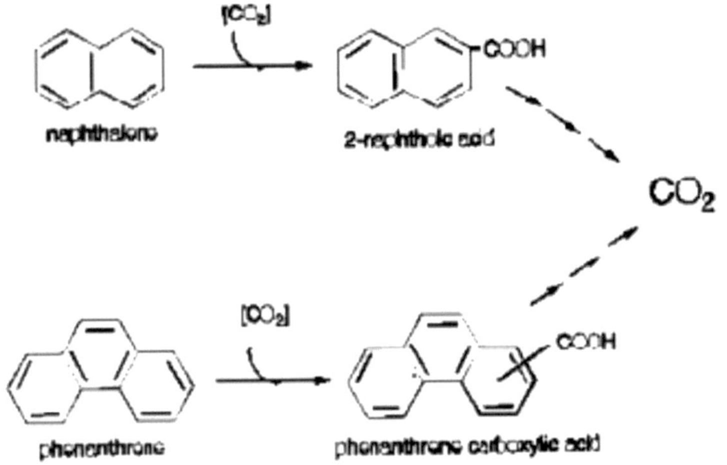

FIGURE 4. Proposed summary pathways for the anaerobic metabolism of NAP and PHE in the sulfidogenic enrichments. Reprinted with permission from Zhang X, Young LY. 1997. Carboxylation as an initial reaction in the anaerobic metabolism of naphthalene and penanthrene by sulfidogenic consortia. Appl Environ Microbiol 63:4759-4764.

From these kinds of experiments, we are able to say that naphthalene, 2-methylnaphthalene, and phenanthrene undergo an initial carboxylation. Remember that there is no oxygen, so the ring must be attacked in a different manner. An initial carboxylation activates the ring and then, through some of our other experiments, we observe a ring reduction that eventually yields ring fission (Zhang and Young 1997; Figure 4).

It may also be helpful to review some of the results we have on anaerobic alkane biodegradation. In a manner similar to the PAHs, we initially used alkane as the sole carbon source and observed whether activity occurred with any of the anaerobic electron acceptors. Again, after fairly long incubation times, we observed activity on octane, decane, and dodecane under sulfate-reducing conditions. In this set of studies, we succeeded in isolating a pure culture of an organism that carries out the reactions. We can now look at it and investigate the mechanism and biochemistry more closely.

We characterized the organism taxonomically and phylogenetically (strain AK-01) as falling within the general class of sulfate reducers. It is a different organism from what has been reported in the literature as an alkane degrader (F. Widdel's group in Bremen, Germany [Aechersberg and others 1998]). We then more closely compared our strain with that from Germany in terms of the degradation mechanisms. Interestingly, they are both sulfate reducers and they are both strict anaerobes that can degrade alkanes, but they attack the alkanes in very different ways. If we used odd-numbered alkanes, C-15 and C-17, AK-01 formed cellular fatty acids that are odd numbered. If we used odd-numbered alkanes for the other organism, the resulting cellular fatty acids were even numbered. The opposite also occurred, i.e., even-numbered alkanes generated even-numbered fatty acids in strain AK-01, and they yielded odd-numbered fatty acids in the other strain. This pattern led us to hypothesize that these two organisms, though cousins, appear to have different mechanisms for alkane degradation. Using substrates that were unlabeled or deuterated or C^{13}-labeled alkanes, we found that the two different strains have very different ways of attacking the alkane under anaerobic conditions.

Strain AK-01 carries out a carbon addition at the subterminal C-2 position of the alkane chain. As a consequence, the terminal carbon then swings down so it forms the methyl group of the C-2 carbon of this fatty acid. Once this occurs, the organism can carry out normal beta-oxidation. It can carry out chain elongation to form larger fatty acids as well. The other strain has a very different attack. It uses inorganic carbonate from solution as the carbon donor. This inorganic carbon is added to the C-3 position of the alkane, and the two terminal carbons are released as acetate. We then end up with two carbons removed and one carbon added to the original alkane so that the resulting fatty acid ends up as an odd-numbered fatty acid. That description is really just the tip of the iceberg because these are only two organisms that have been investigated under anaerobic conditions.

To further contemplate this area of study, consider the following questions:

- Are microorganisms from terrestrial or freshwater systems similar to those found in marine sediments? We know too little about the diversity of these types of degradative anaerobes.
- Are competent organisms actually present? This question is not as straightforward. When we looked for anaerobic toluene and benzene degradation in anoxic-contaminated sediments or in anoxic pristine sediments, our data indicated that the toluene loss and benzene loss occur in the contaminated sediments, but not under the same conditions as in the pristine sediments (Figure 5). This difference suggests that organisms competent for degrading these contaminants are not present in the pristine environments. If they are not there, bioaugmentation may be a viable application for adding organisms where needed.

Let me also point out that there are many pathways for aerobic degradation of PAHs. There are also many aerobic organisms able to carry this out. This kind of diversity has yet to be tapped for the anaerobic microbial community.

We also know about the aerobic degradation of alkanes in terms of

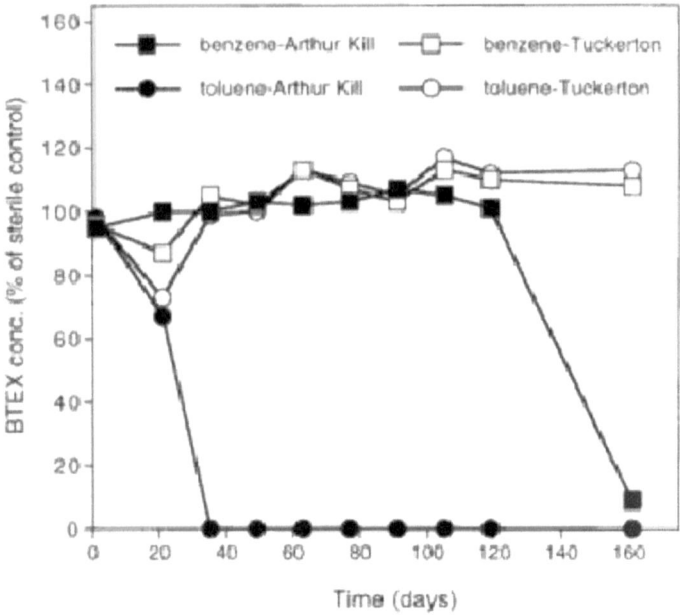

FIGURE 5. Anaerobic biodegradation of BTEX (benzene, toluene, ethylxylene, xylenes) in consortia with sediment from Arthur Kill, New York/New Jersey, harbor (contaminated) and from Tuckerton, New Jersey (uncontaminated). From Phelps CD, Young LY. Unpublished data.

the enzymes and the genes responsible for them. However, there is only one very well characterized aerobic pathway for alkanes. Considering that there are many for the PAHs, there is likely to be others for the alkanes as well; but even in the aerobic realm, our information is limited.

Another issue with respect to the ability of organisms to use different electron acceptors has been addressed by Dr. Jerry Kukor, who studied groundwater from aquifers at three different sites contaminated with benzene, toluene, and xylenes. Data from the three sites indicated that the oxygen was present in significantly lower concentrations compared with the pristine site; furthermore, the levels of oxygen in the contaminant plume is also lower than in the pristine site. In the plume, the potential for both aerobic and anaerobic activity exists. Using the organisms isolated from this site, we can see that low oxygen with nitrate supported much better degradation of toluene than oxygen without nitrate. Here is a hybrid system in which the oxygen is necessary for these organisms because all are aerobic organisms in terms of the mechanisms they use for the degradation process; however, their activity is boosted because they can use nitrate as well as oxygen for respiratory purposes.

I am not sure what mechanism is used for the degradation process, but I would like to add one possible argument for the usefulness of this hybrid. To write an equation describing toluene oxidation to carbon dioxide, 9 mol of oxygen are required for every mol of toluene. To describe it as toluene oxidation first to benzoic acid, aerobically, 1.75 moles of oxygen per mol of toluene as required. This benzoate can then be degraded to carbon dioxide using nitrates and the respiratory electron acceptor. In this case, we have spared the oxygen for the key metabolic activation step. Hence, oxygen is used exclusively for ring activation and nitrate is used for respiration. Thus, microorganisms may have many strategies for biodegradation.

Additional questions to be considered are the following:

- Are any petroleum components inherently recalcitrant so they are not biodegradable?
- Are any petroleum components not bioavailable because of their physical, chemical binding to soils or sediments?
- If the answer above is affirmative, then can regulatory criteria reflect this?
- If petroleum components are not bioavailable, then are they a risk?
- Are anaerobes relevant in remediation of anoxic sediments?
- If the activity of anaerobes is too slow or too minor, do they have a role in cleaning up the environment? In other words, they may not have a significant role in actively cleaning up the environment. Nonetheless, over time, their impact may still be substantial.

If we can get a handle on some of these questions, then we can determine whether certain environments do, indeed, have an intrinsic ability for biodegradation whereas other environments may not.

REFERENCES

Aechersberg F, Rainey FA, Widdel F.
   1998   Growth, natural relationships, cellular fatty acids and metabolic adaptation of sulfate-reducing bacteria that utilize long-chain alkanes under anoxic conditions. Arch Microbiol 170:361-369.

Huntley SL, Bonnevie NL, Wenning RJ.
   1995   Polycyclic aromatic hydrocarbons and petroleum hydrocarbon contamination in sediment from the Newark Bay estuary. Arch Environ Contamin Toxicol 28:93-107.

Sutherland JB, Rafii F, Kahn AA, Cerniglia CE.
   1995   Mechanisms of polycyclic aromatic hydrocarbon degradation. In: Young LY, Cerniglia CE, eds. Microbial Transformation and Degradation of Toxic Organic Chemicals. New York: Wiley-Liss. p 269-306.

Zhang X, Young LY.
   1997   Carboxylation as an initial reaction in the anaerobic metabolism of naphthalene and phenanthrene by sulficogenic consortia. Appl Environ Microbiol 63:4759-4764.

In Situ Bioremediation of Oiled Shoreline Environments

Kenneth Lee

INTRODUCTION

Microbial degradation is a principal process in the elimination of petroleum pollutants from the environment (Cerniglia 1993; Zobell 1964). In consideration of this fact, numerous strategies have been proposed and developed over the last 20 years to accelerate natural oil biodegradation rates. With the reported success of bioremediation operations on the beaches of Alaska after the Exxon Valdez oil spill (Atlas and Bartha 1992; Bragg and others 1994; Prince 1993; Pritchard and Costa 1991), and that of other controlled field trials (Lee and others 1997b; Swannell and others 1996; Venosa and others 1996), this technology is now considered one of the most promising oil spill countermeasures (Hoff 1993; Swannell and Head 1994).

BIOREMEDIATION STRATEGIES

There are two main approaches to oil spill bioremediation: 1) Bioaugmentation involves the addition of oil-degrading bacteria to supplement the existing microbial population; and 2) biostimulation involves the addition of nutrients or growth-enhancing cosubstrates and/or improvements in habitat quality to stimulate the growth of indigenous oil degraders.

Environmental Sciences Division, Maurice Lamontagne Institute, Fisheries and Oceans Canada, Mont-Joli, Quebec, Canada

Bioaugmentation

As a result of extensive media coverage, there is a perception that marine oil spills may be effectively treated by the addition of oil degrading bacteria ("super bugs"). In reality, there is little or no need to add microorganisms to oil-contaminated ecosystems. Microbial ecologists have conclusively demonstrated that oil-degrading bacteria within sediments (Button and others 1992; Lee and Levy 1987; Prince 1993; Venosa and others 1997), open waters (Atlas 1993; Pierce and others 1975), and sea ice (Delille and others 1997) naturally increase in numbers after exposure to oil. Furthermore, field trials have shown that the addition of commercial mixtures (Lee and Levy 1987) or enriched cultures of indigenous oil-degrading bacteria (Fayad and others 1992; Venosa and others 1996) did not significantly enhance the rates of oil biodegradation over that achieved by nutrient enrichment alone. The concept of developing a genetically engineered super bug to degrade crude oil single-handedly is seriously flawed (Lethbridge and others 1994). Vast metabolic potential is required to deal with the diverse array of chemicals in crude oil. Even if it were technically feasible to incorporate all the necessary genetic information into recombinant microorganisms, the burden of maintaining all of these genes is likely to be so great as to make the recombinant strains noncompetitive in the natural environment. In summary, allochthonous microorganisms are generally unable to compete with the natural microflora (Lee and Levy 1987; Venosa and others 1996) in the open environment. Successful enhancement of oil degradation with allochthonous microbial cultures has been achieved only when chemostats or fermentors were used to control conditions and reduce competition from indigenous microflora (Wong and Goldsmith 1988). Although commercial seed cultures may be useful in the treatment of specific compounds within crude oil that are relatively resistant to degradation and isolated spills in confined areas (Lee and Levy 1989a), they appear to be of little benefit for the treatment of the bulk of petroleum contaminants in the open environment. Oil biodegradation within the marine environment is not limited to microbial inocula; therefore, further development of bioremediation agents that contain oil-degrading bacteria as the only active ingredient is difficult to justify.

Biostimulation

Addition of Nutrients

Although the potential capability of indigenous microflora to degrade oil is a function of the physical and chemical properties of the seawater and oil, the environmental conditions, and the biota themselves, it is generally accepted that nutrient availability is the most common limiting factor (Atlas and Bartha 1973; Lee and Levy 1987). Fertilization with nitrogen and phosphorus offers great promise as a countermeasure against marine spills (Atlas and Bartha 1972, 1992; Prince 1993; Swannell and Head 1994; Walker and others 1976) and the ratios of carbon, nitrogen, and phosphorus to support optimal oil degradation rates have been defined (Bragg and others 1994; Reisfeld and others 1972; Venosa and others 1996).

To optimize nutrient delivery, oleophilic nutrient formulations that retain optimal nutrient concentrations at the oil-water interface where biodegradation occurs have been developed (Atlas and Bartha 1973; Tramier and Sirvins 1983). An example is Inipol EAP22 (Elf Aquitaine, France), a microemulsion mixture composed of urea in brine encapsulated in oleic acid as the external phase with lauryl-ether-phosphate as a surfactant (Croft and others 1995; Ladousse and Tramier 1991). Its efficacy has been demonstrated on cobble beaches contaminated by the Exxon Valdez spill in Alaska (Prince 1993). However, additional research on the factors controlling the mechanisms of action is required, as it has not been proven to be effective under all conditions. Failure of bioremediation treatments has been attributed to the rapid loss of nutrients and/or acute toxic responses by the natural microflora to the oil (Lee and Levy 1987; Safferman 1991).

Controlled studies suggest that optimum rates of degradation could be sustained by retaining high, nontoxic, renewable concentrations of nutrients within the interstitial pore water (Lee and others 1997; Venosa and others 1996). The feasibility of adding inorganic nutrients on a periodic basis has been demonstrated in field trials as a means of sustaining elevated nutrient concentrations within the sediments for effective bioremediation (Lee and Levy 1989b, 1991; Venosa and others 1996). The advantages of inorganic agricultural fertilizers as bioremediation agents include low cost, availability, and ease of application.

Field and laboratory beach microcosm studies now suggest that concentrations of nitrate-N for optimal biostimulation should be between 1.0 and 2.5 mg l^{-1} (Bragg and others 1994; Du and others 1999). Although these elevated nutrient concentrations within the interstitial waters in shorelines can be maintained by periodic additions of nutrients, it is not the most practical operational strategy. Nutrient delivery systems must be developed. In this regard, the development of slow-release fertilizer formulations and considerations of beach hydrodynamics in the dispersion of nutrients might decrease cost and effort (Boufadel and others 1999; Lee and others 1993). There is also renewed interest in having an organic carbon source mingled with bioremediation agents to promote rapid bacterial growth (Ladousse and Tramier 1991). This has led to the recent development and testing of organic fertilizers composed of fish meal, animal meal, or fish bone meal (Basseres and others 1993; Lee and others 1995c). Theoretically, optimal nutrient concentrations can be maintained within oiled sediments for prolonged periods by internal nutrient regeneration processes coupled with the degradation of these products, which might also provide essential trace elements and other growth factors.

Addition of Oxygen and Alternate Electron Acceptors

Microbial oil degradation rates within sediments are very slow under anoxic conditions (Atlas and Bartha 1992; Lee and Levy 1991). Sediment tilling and raking have been shown to improve the bioremediation efficacy by increasing the penetration depth of oxygen and nutrient supplements (Sendstad and others 1984; Sergy and others 1998). Although commercial forms of chemical oxidants such as hydrogen, calcium, and magnesium peroxides have been used successfully in terrestrial environments for groundwater remediation, their application in the marine environment warrants further study.

Although carbon transformations by aerobic microorganisms are inhibited in many fine-sediment/wetland environments, facultative and obligate anaerobes become active in anoxic environments and will degrade organic compounds (Patrick and others 1985). Carbon transfer processes in anoxic environments include fermentation, nitrate reduction, denitrification, and sulfate reduction (Valiela 1984). Except for fermentation in which the organic compound itself acts as the terminal electron acceptor, these processes require an inorganic oxidant (e.g., NO_3^- and SO_4^{2-}). Feasibility of bioremediation strategies based on the addition of alternate electron acceptors should be evaluated.

Phytoremediation

Salt marshes are among the most sensitive of ecosystems and the most difficult to clean. Application of traditional oil spill cleanup techniques within this habitat may cause more damage than the oil itself. Foot and mechanical traffic will damage vegetation and drive the hydrocarbons into the anaerobic layer of the sediments where petroleum hydrocarbons may persist for decades (Baker and others 1993). Consideration is now being given to the inherent capacity of wetland plant species to aerate the rhizosphere as a means to stimulate aerobic oil biodegradation. Plants also may take up oil and release exudates and enzymes that stimulate microbial activity. Vegetative transplantation has been used in terrestrial environments for the cleanup of hazardous wastes (Schnoor and others 1995), including polycyclic aromatic hydrocarbons (Banks and

Schwab 1993). Although this process described as phytoremediation has not been used as a marine oil spill countermeasure, recent greenhouse studies with wetland plants (*Spartina* sp.) showed that the oil degradation rate in sediments was significantly enhanced by the application of fertilizer in conjunction with the presence of transplants (Lin and Mendelssohn 1998).

Enhanced Dispersion (Chemical Dispersants, Biosurfactants, Oil-Mineral Fine Interactions)

Microbial attack of oil spilled in the marine environment occurs principally at the oil-water interface. Thus, facilitating an increase in the oil-water interface may enhance the rate and extent of biodegradation as the oil becomes more accessible to nutrients, oxygen, and bacteria. Increases in microbial activity and oil biodegradation have been correlated with the addition of chemical dispersants (Lee and others 1985; Swannell and Daniel 1999), surface agents such as powdered peat (Lee and others 1999), and fertilizers supplemented with biosurfactants for use as bioremediation agents. Research studies after the Exxon Valdez oil spill demonstrated the significance of clay-oil flocculation processes on the natural cleansing of oil residues from impacted shoreline sediment (Bragg and Owens 1994). Physical/chemical interactions with mineral fines reduce the adhesion of the residual oil to sediments by promoting the formation of stable micro-sized oil-fine aggregates (flocs) that are subsequently dispersed into the water column (Bragg and Owens 1994; Lee and others 1997a, 1998). An increase in the oil-water interface facilitated by such oil-mineral fine aggregate formation stimulates both the extent and rate of oil degradation (Lee and others 1997a; Weise and others 1999).

Research during "Spills-of-Opportunity"

In terms of a spill incident case study, the most rigorous study of bioremediation was conducted by Exxon and the US Environmental Protection Agency after the 1989 Exxon Valdez spill in Alaska. Preliminary laboratory experiments demonstrated the potential of nutrient enrichment as a bioremediation treatment (Pritchard and Costa 1991; Pritchard and others 1992). A large-scale (120 km of shoreline in 1989 using 23 tons of nitrogen) field operation was initiated after laboratory and field experiments that confirmed the effectiveness of bioremediation agents that included an oleophilic fertilizer (Bragg and others 1994; Button and others 1992; Glaser and others 1991) dissolved water-soluble (Glaser and others 1991; Pritchard and Costa 1991) and slow-release inorganic fertilizer formulations (Bragg and others 1994; Pritchard and Costa 1991; Safferman

1991), and microbial inocula (Venosa and others 1992). Nutrient treatment was focused on the application of an oleophilic nutrient (Inipol EAP22) for the oil film on surface beach material, and the granular slow-release agricultural fertilizer (Customblen) for subsurface oil. By measuring changes over time in the oil composition relative to hopane, a conserved biomarker, the rate and extent of oil biodegradation was quantified with a high level of statistical confidence. Monitoring hydrocarbon losses relative to this conserved biomarker provided benchmark confirmation of oil biodegradation. Fertilizer additions were reported to accelerate the rate of oil removal by a factor of two to five. Furthermore, it was proven that the rate of oil biodegradation was a function of the nitrogen concentration maintained in the pore water of the intertidal sediment (Bragg and others 1994). These results suggested that the effectiveness of bioremediation can be improved by making real-time measurements of nutrients in sediments to ensure that adequate, but safe, levels of nutrients are maintained during treatment.

In 1996, the Sea Empress grounded at the entrance of Milford Haven, United Kingdom, spilling approximately 65,000 tons of Forties Blend crude oil. Cleanup operations at Amroth Beach after this spill incident provided an opportunity to test the application of surf-washing operations as a means to accelerate the dispersion of oil within the beach sediments into the sea, where it was effectively biodegraded (Lee and others 1997a; Lunel and others 1995) at an enhanced rate.

A RESEARCH NEED FOR OPERATIONAL GUIDELINES

The decision to use bioremediation requires the demonstration of efficacy, reliability, and predictability. Despite successful field demonstrations of its efficacy (Bragg and others 1994; Lee and others 1997b; Prince 1993; Swannell and others 1997; Venosa and others 1996), bioremediation is still a controversial oil spill countermeasure. Part of the problem is that the guidelines for the proper use of the various bioremediation strategies in the marine environment are limited (Swannell and others 1996; Thomas and others 1995). To make informed decisions on the applicability and usage of bioremediation, additional information is required on (1) the testing and selection of bioremediation agents; (2) toxicity and other environmental impacts; (3) the influence of oil chemistry and environmental factors; and (4) the monitoring of efficacy and operational endpoints.

Testing and Selection of Bioremediation Agents

To assist response personnel in the selection and use of spill bioremediation agents, it is useful to have some simple, standard methods for screening performance and toxicity of available bioremediation products (Blenkinsopp and others 1995; Thomas and others 1995).

There is no doubt about the utility of laboratory shaker flask studies to identify the potential impacts and rank the efficacy of various commercial bioremediation agents (Blenkinsopp and others 1995; Pritchard and others 1992; Venosa and others 1997; Wrenn and others 1994). However, laboratory flask studies cannot fully simulate the natural environment where conditions are in a constant state of flux due to tidal cycle inundation and washout, temperature variation, climatic changes, and fresh and saltwater interactions. For example, although ammonium has been used successfully as a nitrogen supplement in field trials (Lee and others 1997b), in small-scale laboratory systems with limited buffering capacity oil biodegradation can be suppressed by acid production associated with ammonia metabolism (Wrenn and others 1994). Indeed, the limitations of both shaker flask and mesocosm tests were recently demonstrated (Lee and others 1997b) as laboratory results could not be reproduced in the field due to physicochemistry changes that altered the interaction between residual oil and sediments.

The need for controlled-release field experiments is evident. Advantages include statistically valid, replicated, randomized block designs with various treatments under conditions that address site heterogeneity and mechanisms of loss.

Different methods have been used to test the efficacy of bioremediation agents in the field. There is now a need for a standard protocol that will allow interlaboratory comparison of results of experiments conducted in different environments (Lee and others 1995a; Merlin 1995). A coordinated effort by the scientific community will accelerate the development of an operational guideline based on a consolidated database of environmentally diverse data.

Toxicity and Other Environmental Impacts

The public has responded favorably to bioremediation strategies based on nutrient enrichment because the implicit goal is that of reducing toxic effects by converting organic molecules to benign cell biomass and "environmentally friendly" products like carbon dioxide and water (Atlas and Cerniglia 1995). Some environmentalists have expressed concern about the net benefit of bioremediation strategies because of the potential production of toxic metabolic by-products, possible toxic components in the formulation of bioremediation agents, and the ineffective degradation of the most toxic components of residual oils (Hoff 1991; OTA 1991). To date, detrimental effects from nutrient enrichment have not been observed after actual field operations (Mearns and others 1997; Prince 1993), although the possibility of a future incident still exists. As an example, oxygen depletion and production of ammonia from excessive applications of a fish-bone meal fertilizer during one field experiment caused detrimental effects that included toxicity and the suppression of oil degradation rates (Lee and others 1995b). For safety assurance, future operational guidelines should include ecotoxicological-monitoring protocols.

DNA analysis may be used to determine population shifts within functional microbial groups as a means to assess stress effects or changes in oil biodegradation potential after bioremediation treatment (Grossman and others 2000). Stable carbon ($\delta^{13}C$) and nitrogen ($\delta^{15}N$) isotopes have been used to monitor changes in trophic interactions after the application of bioremediation agents in the cleanup of oil residues from the Exxon Valdez spill (Coffin and others 1997). Evidence for the transfer of oil-carbon or fertilizer-nitrogen assimilated by bacteria to higher trophic levels has not been found. Assuming bioremediation was effective, additional bacterial biomass arising from oil degradation was either not transferred efficiently to higher trophic levels or not tidally transported from the beach to coastal waters.

Influence of Oil Chemistry and Environmental Factors

A fraction of the components in crude oils spilled within the marine environment are easily degraded; others are slowly or only partially degraded. Some compounds are totally nonbiodegradable (recalcitrant). As a guideline, the greater the complexity (number of alkyl-branched substituents or condensed aromatic rings) of the hydrocarbon structure, the slower the degradation and the greater the likelihood of accumulating partially oxidised intermediary metabolites. These and other factors such as volatility set the practical operational limits of bioremediation strategies. For instance, there is no advantage to bioremediate a surface spill of gasoline because it would evaporate rapidly.

A detailed 7-month study on the bioremediation of a waxy crude oil in sand beach and salt marsh environments has demonstrated the influence of environmental factors on the outcome of a bioremediation treatment strategy (Lee and Levy 1991). Study results clearly demonstrated that the success of bioremediation depends on the nature of the contaminated shoreline. On a sandy beach contaminated with low concentrations of Terra Nova crude oil, toxicity to the oil-degrading bacteria was not a factor, and ambient concentrations of nitrogen and phosphorus were sufficient to result in rapid oil biodegradation. Under these conditions, nutrient enrichment provided little or no benefit and nature can be left to take its course (a nonaction strategy). However, higher oil levels provided a carbon-enriched environment and the microbial community within the beach became nutrient-limited, and bioremediation treatment could effectively enhance the rate of oil removal. In the salt marsh environment treated with similar oil concentrations, oil penetrated into the anoxic layers of the sediment and the fertilization strategy was ineffective. In this particular case, the addition of oxygen may be required as a part of the bioremediation strategy. The intricacy of interactions influencing the success of bioremediation in this study is not unique. The ability of indigenous microbes of Prince William Sound, Alaska (Sugai and others 1997), to mineralize hexadecane, phenanthrene, and naphthalene has been shown to be influenced by the intensity of physical mixing, the method of bioremediation agent application, and the availability of alternative carbon sources.

The efficacy of specific bioremediation formulations may be influenced by environmental conditions. For example, at temperate conditions greater than 15° C, slow-release (sulphur-coated urea) fertilizer formulations appear to be more effective in retaining elevated nutrient concentrations within the sediments than inorganic nitrogen (ammonium nitrate) fertilizers (Lee and others 1993). Lower temperatures are thought to reduce the permeability of the coating on the slow-release fertilizer, effectively suppressing nutrient release rates. For optimal effectiveness, the selection of bioremediation agents should take into account the environmental conditions, the type of contaminated shoreline, and the methods of application (Lee and others 1993; Prince 1993; Swannell and others 1995, 1996).

Studies in the intertidal region of sandy beaches with lithium as a conservative tracer (Wrenn and others 1997) have demonstrated that dissolved nutrient transport is driven by tide-influenced hydraulic gradients and wave activity. Nutrient retention in the bioremediation zone of sand beach could be predicted from data on the extent of water coverage, and a suitable application schedule could be devised from the modeling of hydrodynamic data.

In north-temperate environments, although winter temperatures do not affect the apparent number of heterotrophic bacteria in oiled sediments, the number of oil-degraders declines (Lee and Levy 1989b, 1991; Prince 1993; Swannell and others 1997). Further study is warranted to identify whether these observations are attributed to a physiological response or to physiochemical changes in the oil that alters its availability to the bacteria. It is now also apparent that the most important influence on the carrying capacity for hydrocarbon degraders in the marine environment may be the removal of biomass by physical processes such as scouring by breaking waves. If this is the case, the optimal level of oil degradation capacity can be provided by indigenous bacteria provided that sufficient nutrients are present. The addition of exogenous hydrocarbon degraders (i.e., bioaugmentation) will not increase population density (Venosa and others 1996).

Rapid biodegradation of crude oil stranded within intertidal environments can occur under temperate conditions. On the Delaware coast, natural nitrogen concentrations were found to be high enough to sustain rapid intrinsic rates of biodegradation without human intervention (Venosa and others 1996). Although nutrient addition at this site significantly accelerated the rate of hydrocarbon biodegradation, the incremental increase (slightly greater than 200% for the alkanes and 50% for the polynuclear aromatic hydrocarbon levels) is not high enough to warrant a major, perhaps costly, bioremediation effort in the event of a large crude oil spill in that area. A similar conclusion was also reached in a field trial to evaluate the influence of a slow-release fertilizer on the biodegradation rate of crude oil spilled on intertidal sediments of an estuary (Oudot and others 1998). Due to adaptation of marine bacteria to hydrocarbons along the coast of Brittany (Atlas and Cerniglia 1995) and high background levels of N and P at the study site, no significant difference in biodegradation rates was detected after nutrient addition. It was proposed that bioremediation by nutrient enrichment would be of limited use if background interstitial porewater levels of N exceed 100 µmoles l^{-1}. A strong correlation between the available concentrations of ammonia and phosphorus and the degradation rates of petroleum has been demonstrated in a recent study in Texas that monitored the relatively rapid recovery of an oil-impacted coastal wetland environment by intrinsic biodegradation (Harris and others 1999). In light of these results, it is suggested that interstitial nutrient levels be determined before any decision is made to apply bioremediation agents.

Monitoring Remediation Effectiveness and Identification of Operational Endpoints

Wide acceptance and use of bioremediation strategies by the oil spill response community has been limited by the lack of defined performance standards. For proper application of the technology, there is a need for monitoring programs to quantify intrinsic rates of oil loss and degradation, demonstrate treatment efficacy, and identify operational endpoints.

A major obstacle is heterogeneity within the natural environment. Absolute levels of contamination can vary widely over a site and simple estimates of biodegradation based on sequential samples can be confounded by this heterogeneity, unless large numbers of samples are taken. This problem can be resolved by the normalization of data to conserved markers such as hopanes and chrysenes found within the oil (Lee and others 1997b; Oudot and others 1998; Prince and others 1993; Venosa and others 1996). Though costly and time-consuming, these analyses by gas chromatography/mass spectrometry are necessary to demonstrate effectiveness at a level of precision and accuracy demanded by the scientific community. However, from an operational perspective, considering the numerous samples needed to characterize a spill site, other more rapid and less costly performance measures must be developed to satisfy regulators and managers.

In situ measurement of microbial CO_2 production by respirometry or radiotracer methods can be used to quantify oil mineralisation rates to estimate bioremediation success (Swannell and others 1994, 1997). Enumeration of potential oil-degrading bacteria by their isolation on specific media has become a benchmark in many bioremediation studies, although many bacteria within the natural environment are dormant or unculturable on the media used. Therefore, it is essential to show, by combined chemical and microbiological methods, that the oil-degrading bacteria are truly active.

Recent studies have shown changes in the distribution of hydrocarbon-degrading genes in response to the hydrocarbon composition to which the bacterial population is exposed (Sotsky and others 1990). Future use of DNA and RNA gene probes for pollutant catabolic pathways may provide practical and evolutionary insights into how and why biodegradation activity is expressed (Greer and others 1993; Sayler and Layton 1990).

As discussed, future operational guidelines will incorporate reliable microbial response and ecotoxicological monitoring protocols to verify efficacy for toxicity reduction over that of no treatment. In addition to direct chemical evidence of oil degradation, microscale biotests may provide an operational endpoint indicator for bioremediation on the basis of toxicity reduction; i.e., the site is acceptable as there is no detectable toxic effects, or the treatment is detrimental in that a toxic response is induced (Lee and others 1995b; Mearns and others 1995).

CONCLUSIONS

With the recent demonstrations of its efficacy in the field, bioremediation has been touted as the emerging oil spill countermeasure of the 21st century. An advantage of this environmentally friendly technology is its relatively low cost, as it does not require large numbers of personnel or highly specialized equipment for its application. However, its wide acceptance as an operational oil spill countermeasure has been limited by the lack of data showing its effectiveness relative to current technologies and operational guidelines for its application.

Operational limitations exist for all oil spill countermeasures. In the context of shoreline cleanup, bioremediation should be considered a useful addition to the toolbox of oil spill treatment strategies, including the option of "no treatment." Improvements in bioremediation technologies will result from basic research in microbial ecology, which will identify the factors controlling optimal rates of oil degradation. Future applied research is also needed to construct a database for decision making that includes information on the type of oil, application methodologies available (form and type of bioremediation agent, type and frequency of application), environmental conditions (availability of nutrients, bacteria, oxygen, temperature, and wave or tidal immersion), and defining treatment endpoints.

REFERENCES

Atlas RM.
   1993   Bacteria and bioremediation of oil spills. Oceanus 36:71-73.
Atlas RM, Bartha R.
   1992   Hydrocarbon biodegradation and oil spill bioremediation. Adv Microb Ecol 12:287-338.
Atlas RM, Bartha R.
   1972   Degradation and mineralization of petroleum in seawater: Limitations by nitrogen and phosphorous. Biotechnol Bioeng 14:309-317.
Atlas RM, Bartha R.
   1973   Stimulated biodegradation of oil slicks using oleophilic fertilizers. Environ Sci Technol 7:538-541.
Atlas RM, Cerniglia CE.
   1995   Bioremediation of petroleum pollutants. Bioscience 45:332-338.
Baker JM, Guzman LM, Bartlett PD, Little DI, Wilson CM.
   1993   Long-term fate and effects of untreated thick oil deposits on salt marshes. In: Proceedings of the International Oil Spill Conference. Washington, DC: American Petroleum Institute. p 395-399.
Banks KM, Schwab AP.
   1993   Dissipation of polycyclic aromatic hydrocarbons in the rhizosphere. In: Symposium on Bioremediation of Hazardous Wastes: Research, Development and Field Evaluations, Washington, DC: Environmental Protection Agency, EPA/600/R-93/054, p 246.
Basseres A, Eyraud P, Ladousse A, Tramier B.
   1993   Enhancement of spilled oil biodegradation by nutrients of natural origin. In: Proceedings of the International Oil Spill Conference. Washington, DC: American Petroleum Institute. p 495-501.
Blenkinsopp S, Sergy G, Wang Z, Fingas MF, Foght J, and Westlake DWS.
   1995   Oil spill bioremediation agents—Canadian efficacy test protocols 1995. In: Proceedings of the 1995 International Oil Spill Conference. Washington, DC: American Petroleum Institute. p 91-96.
Boufadel MC, Suidan MT, Rauch CH, Ahn C-H, Venosa AD.
   1999   Nutrient transport in beaches subjected to freshwater input and tides. In: Proceedings of the International Oil Spill Conference. Washington, DC: American Petroleum Institute. (Publication 4686A, Paper 170).

Bragg JR, Owens EH.
  1994   Clay-oil flocculation as a natural cleansing process after oil spills: Part 1: Studies of shoreline sediments and residues from past spills. In: Proceedings of the 17th Arctic and Marine Oil Spill Program (AMOP) Technical Seminar, Vancouver, British Columbia. p 1-24.

Bragg JR, Prince RC, Harner EJ, Atlas RM.
  1994   Effectiveness of bioremediation for the Exxon Valdez oil spill. Nature 368:413-418.

Button DK, Robertson BR, McIntosh D, Juttner F.
  1992   Interactions between marine bacteria and dissolved-phase and beached hydrocarbons after the Exxon Valdez oil spill. Appl Environ Microbiol 58:243-251.

Cerniglia CE.
  1993   Biodegradation of polycyclic aromatic hydrocarbons. Curr Opin Biotechnol 4:331-338.

Coffin RB, Cifuentes LA, Pritchard PH.
  1997   Assimilation of oil-derived carbon and remedial nitrogen applications by intertidal food chains on a contaminated beach in Prince William Sound, Alaska. Mar Environ Res 44:27-39.

Croft B, Swannell RPJ, Grant AL, Lee K.
  1995   Effect of bioremediation agents on oil biodegradation in medium-fine sand. In: Hinchee RE, Kittel JA, Reisinger HJ, eds. Applied Bioremediation of Petroleum Hydrocarbons. Bioremediation. Vol 3. Columbus, OH: Battelle Press. p 423-434.

Delille D, Basseres A, Dessommes A.
  1997   Seasonal variation of bacteria in sea ice contaminated by diesel fuel and dispersed fuel oil. Microb Ecol 33:97-105.

Du X, Resser P, Suidan MT, Huang T, Moteleb M, Boufadel MC.
  1999   Optimum nitrogen concentration supporting maximum crude oil biodegradation in microcosms. In: Proceedings of the International Oil Spill Conference. Washington, DC: American Petroleum Institute. (Publication 4686A, Paper 222).

Fayad NM, Edora RL, El-Mubarak AH, Polancos AB.
  1992   Effectiveness of a bioremediation product in degrading the oil spilled in the 1991 Arabian Gulf war. Bull Environ Contam Toxicol 49:787-796.

Glaser JA, Venosa AD, Opatken J.
  1991   Development and evaluation of application techniques for delivery of nutrients to contaminated shoreline in Prince William Sound. In: Proceedings of the International Oil Spill Conference. Washington, DC: American Petroleum Institute. p 559-562.

Greer CL, Masson Y, Comeau R, Brousseau R, Samson R.
  1993   Application of molecular biology techniques for isolating and monitoring pollutant degrading bacteria. Water Poll Res J Canada 28:275-287.

Grossman M, Prince R, Garrett R, Garrett K, Bare R, Lee K, Gergy G, Owens E, Guenette C.
  2000   Microbial diversity in oiled and unoiled shoreline sediments in the Norwegian Arctic. In: Bell CR, Brylinsky M, Johnson-Green P, eds. Proceedings of the 8th International Symposium on Microbial Ecology (ISME-8), Halifax, Nova Scotia, August 9-14, 1998. Atlantic Canada Society for Microbiological Ecology. p 775-789.

Harris BC, Bonner JS, Autenrieth RL.
  1999   Nutrient dynamics in marsh sediments contaminated by an oil spill after a flood. Environ Technol 20:795-810.

Hoff R.
  1991   A summary of bioremediation applications observed at marine oil spills. Report HMRB 91-2. Washington, DC: Hazardous Materials Response Branch, National Oceanic and Atmospheric Administration. 30 p.

Hoff R.
1993　Bioremediation: An overview of its development and use for oil spill cleanup. Mar Pollut Bull 26:476-481.
Ladousse A, Tramier B.
1991　Results of 12 years of research in spilled oil bioremediation: INIPOL EAP22. In: Proceedings of the International Oil Spill Conference. Washington, DC: American Petroleum Institute. p 577-581.
Lee K, Levy EM.
1987　Enhanced biodegradation of a light crude oil in sandy beaches. In: Proceedings of the International Oil Spill Conference. Washington, DC: American Petroleum Institute. p 411-416.
Lee K, Levy EM.
1989a　Biodegradation of petroleum in the marine environment and its enhancement. In: Nriagu JA, Lakshminarayana JSS, eds. Aquatic Toxicology and Water Quality Management. New York: Wiley & Sons Inc. p 217-243.
Lee K, Levy EM.
1989b　Enhancement of the natural biodegradation of condensate and crude oil on beaches of Atlantic Canada. In: Proceedings of the International Oil Spill Conference. Washington, DC: American Petroleum Institute. p 479-486.
Lee K, Levy EM.
1991　Bioremediation: Waxy crude oils stranded on low-energy shorelines. In: Proceedings of the International Oil Spill Conference. Washington, DC: American Petroleum Institute. p 541-547.
Lee K, Merlin FX, Swannell RPJ, Reilly T, Sveum P, Oudot J, Guillerme M, Ducreuz J, Chaumery C.
1995a　A protocol for experimental assessments of bioremediation strategies on shorelines. In: Proceedings of the International Oil Spill Conference. Washington, DC: American Petroleum Institute. p 901-902.
Lee K, Siron R, Tremblay GH.
1995b　Effectiveness of bioremediation in reducing toxicity in oiled intertidal sediments In: Hinchee RE, Brockman FJ, Vogel CM, eds. Microbial Processes for Bioremediation. Columbus, OH: Battelle Press, p 117-128.
Lee K, Tremblay GH, Cobanli SE.
1995c　Bioremediation of oiled-beach sediments: Assessment of inorganic and organic fertilizers. In: Proceedings of the International Oil Spill Conference. Washington, DC: American Petroleum Institute. p 107-113.
Lee K, Lunel T, Wood P, Swannell R, Stoffyn-Egli P.
1997a　Shoreline cleanup by acceleration of clay oil-flocculation processes. In: Proceedings of the International Oil Spill Conference. Washington, DC: American Petroleum Institute. p 235-240.
Lee K, Tremblay GH, Gauthier J, Cobanli SE, Griffin M.
1997b　Bioaugmentation and biostimulation: A paradox between laboratory and field results. In: Proceedings of the International Oil Spill Conference. Washington, DC: American Petroleum Institute. p 697-704.
Lee K, Stoffyn-Egli P, Wood P, Lunel T.
1998　Formation and structure of oil-mineral fine aggregates in coastal environments. In: Proceedings of the 21st Arctic and Marine Oil Spill Program (AMOP) Technical Seminar, June 10-12, 1998, Edmonton, Alberta. p 911-921.
Lee K, Cobanli SE, Gauthier J, St.-Pierre S, Tremblay GH, Wohlgeschaffen GD.
1999　Evaluating the addition of fine particles to enhance oil degradation. In: Proceedings of the International Oil Spill Conference. Washington, DC: American Petroleum Institute. (Publication 4686A, Paper 433).

Lee K, Tremblay GH, Levy EM.
  1993 Bioremediation: Application of slow-release fertilizers on low-energy shorelines. In: Proceedings of the International Oil Spill Conference. Washington, DC: American Petroleum Institute. p 449-454.

Lee K, Wong CS, Cretney WJ, Whitney FA, Parsons TR, Lalli CM, Wu J.
  1985 Microbial response to crude oil and Corexit 9527: Seafluxes enclosure study. Microb Ecol 11:337-351.

Lethbridge G, Vits HJJ, Watkinson RJ.
  1994 Exxon Valdez and bioremediation. Nature 371:97-98.

Lin Q, Mendelssohn IA.
  1998 The combined effects of phytoremediation and biostimulation in habitat restoration and enhancement of oil degradation for petroleum-contaminated wetlands. Ecol Eng 10:263-274.

Lunel T, Swannell RPJ, Rusin J, Wood P, Bailey N, Halliwell C, Davies L, Sommerville M, Dobie A, Mitchell D, McDonagh M, Lee K.
  1995 Monitoring the effectiveness of response operations during the Sea Empress incident: A key component of the successful counter-pollution response. Spill Sci Technol 2:99-112.

Mearns AJ, Doe K, Fisher W, Hoff R, Lee K, Siron R, Mueller C, Venosa A.
  1995 Toxicity trends during an oil spill bioremediation experiment on a sandy shoreline in Delaware, USA. In: Proceedings of the 18th Arctic and Marine Oil Spill Program (AMOP) Technical Seminar, Environment Canada, p 1133-1145.

Mearns AJ, Venosa AD, Lee K, Salazar M.
  1997 Field testing bioremediation treating agents: lessons learned from an experimental shoreline oil spill. In: Proceedings of the International Oil Spill Conference. Washington, DC: American Petroleum Institute. p 707-712.

Merlin FX.
  1995 Devising an experimental protocol to evaluate the effectiveness of bioremediation procedures. In: Proceedings of the 2nd International Oil Spill Research and Development Forum. London: International Maritime Organization. p 37-44.

OTA [US Office of Technology Assessment].
  1991 Bioremediation of Marine Oil Spills - Background Paper OTA-BP-O-70. Washington, DC: GPO. p 31.

Oudot J, Merlin FX, Pinvidic P.
  1998 Weathering rates of oil components in a bioremediation experiment in estuarine sediments. Mar Environ Res 45:113-125.

Patrick WH, Mikkelsen DS, Wells BR.
  1985 Plant nutrient behavior in flooded soil. In: Engelstad OP, ed. Fertilizer Technology and Use. Madison, WI: Soil Science Society of America. p 197-228.

Pierce RH, Cundell AM, Traxler RW.
  1975 Persistence and biodegradation of spilled residual fuel oil on an estuarine beach. Appl Microbiol 29:646-652.

Prince RC.
  1993 Petroleum spill bioremediation in marine environments. Crit Rev Microbiol 19: 217-242.

Prince RC, Drake EN, Madden PC, Douglas GS.
  1993 Biodegradation of polycyclic aromatic hydrocarbons in a historically contaminated soil. In: Polycyclic Aromatic Hydrocarbons, In-Site and On-Site Bioremediation. Vol 2. p 205-210.

Pritchard PH, Costa CF.
  1991 EPA's Alaska Oil Spill Bioremediation Project. Environ Sci Technol 25:372-379.

Pritchard PH, Mueller JG, Rogers JC, Kremer FV, Glaser JA.
  1992 Oil spill bioremediation: experiences, lessons, and results from the Exxon Valdez oil spill in Alaska. Biodegradation 3:315-335.
Reisfeld A, Rosenberg E, Gutnick D.
  1972 Microbial degradation of crude oil: Factors affecting the dispersion in seawater by mixed and pure cultures. Appl Microbiol 24:363-368.
Safferman SI.
  1991 Selection of nutrients to enhance biodegradation for the remediation of oil spilled on beaches. In: Proceedings of the International Oil Spill Conference. Washington, DC: American Petroleum Institute. p 571-576.
Sayler GS, Layton AC.
  1990 Environmental application of nucleic acid hybridization. Ann Rev Microbiol 44:625-648.
Schnoor JL, Licht LA, McCutcheon SC, Wolfe NL, Carreira LH.
  1995 Phytoremediation of organic and nutrient contaminants. Environ Sci Technol 29:318-323.
Sendstad E, Sveum P, Endal LJ, Brattbakk Y, Ronning O.
  1984 Studies on a seven year old seashore crude oil spill on Spitsbergen. In: Proceedings of the 7th Arctic and Marine Oil Spill Program (AMOP) Technical Seminar, Environment Canada. p 60-74.
Sergy G, Guénette C, Owens E, Prince RC, Lee K.
  1998 The Svalbard shoreline experimental oil spill field trials. In: Proceedings of the 21st Arctic and Marine Oilspill Program (AMOP) Technical Seminar. June 10-12, 1998, Edmonton, Alberta. p 873-889.
Sotsky JB, Greer CW, Atlas RM.
  1990 Frequency of genes in aromatic and aliphatic hydrocarbon biodegradation pathways within bacterial populations from Alaskan sediments. Can J Microbiol 40:981-985.
Sugai SF, Lindstrom JE, Braddock JF.
  1997 Environmental influences on the microbial degradation of Exxon Valdez oil on the shorelines of Prince William Sound, Alaska. Environ Sci Technol 31:1564-1572.
Swannell RPJ, Basseres A, Lee K, Merlin FX.
  1994 A direct respirometric method for the in situ determination of bioremediation efficacy. In: Proceedings of the 17th Arctic and Marine Oilspill Program (AMOP) Technical Seminar, Environment Canada. p 1273-1286.
Swannell RPJ, Croft BC, Grant AL, Lee K.
  1995 Evaluation of bioremediation agents in beach microcosms. Spill Sci Technol 2:151-159.
Swannell RPJ, Daniel F.
  1999 Effect of dispersants on oil biodegradation under simulated marine conditions. In: Proceedings of the International Oil Spill Conference. Washington, DC: American Petroleum Institute. (Publication 4686A, Paper 212).
Swannell RPJ, Head IM.
  1994 Bioremediation comes of age. Nature 368:396-397.
Swannell RPJ, Lee K, McDonagh M.
  1996 Field evaluations of marine oil spill bioremediation. Microbiol Rev 60:342-365.
Swannell RPJ, Mitchell DJ, Jones DM, Willis AL, Lee K, Lepo JE.
  1997 Field evaluation of bioremediation to treat crude oil in a mudflat. In: In situ and on-site bioremediation. Proceedings of the 4th International In Situ and On-site Bioreclamation Symposium. p 401-406.

Thomas G, Nadeau R, Ryabik J.
  1995   Increasing readiness to use bioremediation in response to oil spills. In: Proceedings of the 2nd International Oil Spill Research and Development Forum. London: International Maritime Organization. p 56-62.
Tramier B, Sirvins A.
  1983   Enhanced oil biodegradation: A new operational tool to control oil spills. In: Proceedings of the 1983 Oil Spill Conference. Washington, DC: American Petroleum Institute. p 115-119.
Valiela I.
  1984   Marine Ecological Processes. New York: Springer-Verlag. 546 p.
Venosa AD, Haines JR, Eberhart BL.
  1997   Screening of bacterial products for their crude oil biodegradation effectiveness. Methods in Biotechnology, 2. In: Sheehan D, ed. Bioremediation Protocols. Totowa, NJ: Humana Press Inc. p 47-58.
Venosa AD, Haines JR, Nisamaneepong W, Govind WR, Pradhan S, Siddique B.
  1992   Efficacy of commercial products in enhancing oil biodegradation in closed laboratory reactors. J Ind Microbiol 10:13-23.
Venosa AD, Suidan MT, Wrenn BA, Strohmeier KL, Haines JR, Eberhart BL, King D, Holder EL.
  1996   Bioremediation of an experimental oil spill on the shoreline of Delaware Bay. Environ Sci Technol 30:1764-1775.
Walker JD, Colwell RR, Petrakis L.
  1976   Biodegradation rates of components of petroleum. Can J Microbiol, 22:1209-1213.
Weise AM, Nalewajko C, Lee K.
  1999   Oil-mineral fine interactions facilitate enhanced oil biodegradation in seawater. Environ Technol 20:811-824.
Wong AD, Goldsmith CD.
  1988   The impact of a chemostat discharge containing oil degrading bacteria on the biological kinetics of a refinery activated sludge process. Water Sci Technol 20:131-136.
Wrenn BA, Haines JR, Venosa AD, Kadhodayan M, Suidan MT.
  1994   Effects of nitrogen source on crude oil biodegradation. J Ind Microbiol 13:279-286.
Wrenn BA, Suidan MT, Strohmeier KL, Eberhardt BL, Wilson GJ, Venosa AD.
  1997   Nutrient transport during bioremediation of contaminated beaches: Evaluation with lithium as a conservative tracer. Water Res 31:515-524.
Zobell CE.
  1964   The occurrence, effects, and fate of oil polluting the sea. Adv Water Pollut Res 3:85-119.

Contributions of Marine Biotechnology to Marsh Oil Spill Restoration

Ralph J. Portier

INTRODUCTION

There is an estimated 3.2 million tons annual (mta) input of petroleum hydrocarbons into the world's oceans (NRC 1985). The majority is in small amounts from chronic sources, 0.7 mta from tanker operations, and 0.7 mta from municipal wastes. Accidental spills account for 0.42 mta, just 13% of the world's total input of petroleum hydrocarbons. The chronic, small amounts of oil are rapidly removed from the marine environment by a variety of processes—evaporation, dissolution, biodegradation, emulsification, and sedimentation—in a matter of days in normal conditions. When there is an accidental spill from oil production or transport leading to a large lens of visible brown/black oil, the environment's natural capacity for self-purification is overwhelmed. The oil may persist for months if not decades. Serious acute and chronic ecological damage can occur, and economies and community health can be affected (Atlas and Bartha 1973; Kelso and Kendziorek 1991; Overton and others 1994). Because of the danger to health, ecology, and public relations represented by large oil spills that overwhelm natural capacity for purification, new marine biotechnology approaches are needed to move the "technology" forward for cleaning up impacted coastal and marsh environments.

The fate of petroleum hydrocarbons in the marine environment has been documented by Bartha (1986). A small oil spill will spread out until

---

Aquatic/Industrial Toxicology Laboratory, Institute for Environmental Studies, Louisiana State University, Baton Rouge, LA

it is just a sheen on the water surface; 1 g will cover 1-10 m^2. This thin film will be evaporated, emulsified, metabolized, or dissolved. Depending on temperature, mixing conditions and composition of the oil, 10-55% will be lost through evaporation and photo-oxidation (Baker and others 1993; Walker and others 1993). The more polar fractions of the oil, carbon lengths 12 and less ($\leq$ C-12), will dissolve, ultimately to be metabolized by naturally occurring bacteria (Overton and others 1994). Natural processes will emulsify the remaining oil or it will have an impact on the sea bottom or marsh environment. If the oil undergoes emulsification and natural dispersion, then within 2 months, the bioavailable hydrocarbons will be metabolized, leaving behind a highly condensed, recalcitrant residue of complex hydrocarbons called asphaltenes and resins (Bartha 1986; Stewart and others 1993).

If conditions are poor for emulsification and dispersion of the oil, typical for marsh environments, it may emulsify only partly, forming a mousse, which is an oil-in-water emulsion (up to 80% water, depending on the oil) that is highly resistant to degradation. Mousse has been known to persist in sediments for decades (Atlas 1981; Baker and others 1993; Bartha 1986; NRC 1985).

OIL SPILL EFFECTS

Oil spills affect ecosystems in three ways: smothering plants and animals, massive input of organic carbon upsetting nutrient cycling, and toxicity (NRC 1985).

- **Smothering.** Smothering of plants and animals comes about due to oil's physical characteristics—its stickiness, buoyancy, and oleophilicity.
- **Disruption of nutrient cycling and microbial diversity.** The normal nutrient cycle will be disrupted by the massive influx of hydrocarbon. This will exert a selective pressure on the microbial biota for petroleum hydrocarbon degradation (Bartha 1986). This selection pressure will change the natural biodiversity, perhaps changing the flow of energy through the marine food web and ultimately changing what food sources are available to higher organisms.
- **Toxicity.** Oil exerts its toxic effects primarily through its water-soluble fractions. Hydrophobic fractions will exert toxic effects only if swallowed or adhered to the skin where hydrophobic compounds can dissolve into lipophilic tissues. The water-soluble fractions are more toxic because they dissolve in the water, thus coming into contact with marine biota not near the oil spill. As the more complex and less soluble compounds are oxidized in metabolism and photo-oxidation, they become water soluble and begin to affect the biota. Effects seen with toxic hydro-

carbon and hydrocarbon residues are changes in respiration, growth, reproduction, behavior, calcification, molting, ion transport, and enzyme activity (NRC 1985).

RESPONSE AND LIMITATIONS

Oil spill response aims to prevent damaging effects by removing the oil from the endangered environment. A variety of spill-response methods exist and are generally broken down into two classes:

- **Mechanical response.** Mechanical response at sea is the use of booms and other physical devices to contain and aid in physical recovery of the oil. This method has rarely been used to its full theoretical capability due to bad weather, sea state, or logistical problems related to the volume of oil spilled in a catastrophic accident.
- **Chemical response.** Chemical response to oil spills at sea consists of applying dispersants to disperse the oil as tiny droplets into the water. This was used to great effect in the spill from the Sea Empress off the coast of Wales in February 1966 (Lunel and others 1997). Some success has also been achieved with surfactant beach cleaners that are designed to lift oil from beaches without dispersing it (Prince and others 1999).

However, there was and continues to be concern over the combined effect of oil and dispersants (George-Ares and others 1999; Wolfe and others 1998). Although dispersants are no longer more toxic than the oil they are supposed to remediate, they will increase the toxic effect of the oil. As stated above, it is primarily the water-soluble fraction of the oil that is toxic because of its transport through water to the organism. For the normal oil slick on the marsh surface, only the organisms near the air/water interface of the oil will encounter high concentrations of toxins. When the oil has been dissolved into the water column, as happens with dispersants, deep water biota not normally affected by oil spills will encounter oil. The current thinking on spill response to coastal marine environments is summarized in Table 1.

TABLE 1. Current Remediation Approaches: Marsh Habitat

1. Boom It!
2. Disperse It!
3. Floc It!
4. Burn It!
5. Bug It!
6. Ah....... Just Forget It!

MARINE BIOTECHNOLOGY CONTRIBUTIONS

Biologicals

The development of commercial inocula for industrial wastewater biotreatment is a mature industry. Microbial products are used daily by coastal zone industries to treat elevated wastewater discharges into littoral environments. Most of these products are adapted microflora packaged on a pasteurized wheat bran base. Minimal toxicological testing of these products has been conducted to date. Similar products proposed for use in oil spill response in these coastal environments have undergone a comprehensive series of tiered tests under federal guidelines (Portier 1991). Few products have been approved to date for US Coast Guard use in impacted marsh environments. Biologicals include the aforementioned whole cell products, enzyme preparations, co-oxidizing substrates, modifying agents, and nutrient amendments. There is a need to further expand the type, efficacy, and total number of such products available for marsh restoration. Critical needs for additional research are summarized in Table 2.

Engineered Systems for Marsh Habitat

With the development of a more efficacious battery of biologicals, engineered systems that deliver the novel biotech product with precision and minimal impact are also needed. Current protocols for delivering biologicals are rather primitive. Mechanical sprayers are the current state of the art. Engineered systems are needed for preinvasive response to oiling and post-oil ablation. A robust screening protocol to test candidate engineered systems must be developed for the unique marsh habitat. Engineered systems approved for a neritic/pelagic environment may not be appropriate for the littoral environment. Positioning equipment that delivers biological and/or combination products with minimal marsh impact are still needed. Finally, spill response companies must be weaned off expensive, lucrative, but hopelessly ineffective booming and chemical treatment strategies (Portier and Ahmed 1988).

TABLE 2. Marsh Habitat: Biologicals Development

1. Need for a "better bug"
2. Improvement of dynamics for indigenous populations to perform
3. Can enzyme preparations be used?
4. Unique enzymology from acclimated mesophiles: availability and efficacy
5. Licensing and risk assessment of novel biologicals

Analytical Approaches: "Real Time" Aids to Remediation

If a better biological coupled to an acceptable engineered system can be realized from marine biotechnology research, the question one must then pose is "How can we assess the efficacy of treatment?" There has always been a linkage between spill response and analytical instrumentation. Traditional gas chromatography/mass spectroscopy protocols have been developed for assessing and fingerprinting oil, yet the instrumentation is bulky and not relatively mobile. A prototype portable device is under final field testing and will be available in 2000 (Overton and others 1994). However, this device is really the first of a new generation of handheld sophisticated tools for assessing impact from a spill. A summary of analytical instrumentation development linked to marine biotechnology research programs appears in Table 3.

Development of Risk Assessment Strategies for Marsh Habitats

Finally, there still is a need to predict risk and relative impact. Assuming logistics and intervention approaches have become more sophisticated through the years, there continues to be the problem of developing the environmental management tools to determine when and if a marine biotechnology delivery system will minimize and/or facilitate postspill remediation (Portier and Ahmed 1988; Smith and Portier 1997). Biological assays are effective tools in assessing impact from point-source wastewater discharges or from impacted soils. Few assays are available for assessing acute and chronic toxicity of benthic and marsh habitat. A battery of sophisticated, possibly genome-based, assays need to be developed for marsh grasses, marsh mammalian populations, microorganisms, and crustacea (Lee and Portier 1999; Lin and others 1999).

CONCLUSIONS

Marine biotechnology approaches can play a pivotal role in developing strategies for prevention and/or postevent restoration of marsh habitats. The focus for the past few decades has been on crude oil and refined petroleum products. Domestic sewage and small volume-generated point

TABLE 3. Analytical Instrumentation for Marsh Restoration

1. Real time instrumentation
2. Instrumentation focused on benthic and plant biota
3. Instrumentation to measure toxicological impact
4. Instrumentation to measure chronic effects

sources pose greater threats to the coastal marsh environment annually. Thus, new tools will be needed to assess, model, prevent, and restore spills in our nation's coastal zone. To summarize, the following actions should be considered for fundamental research in marsh restoration:

- Establish linkages to existing National Science Foundation centers to further develop novel biologicals for spill response.
- Establish a program review on biotechnology products/engineered systems assessment and approval for field applications.
- Continue to look for low-tech or "no" tech approaches based on risk assessment strategies.

REFERENCES

Atlas R.
　1981　Microbial degradation of petroleum hydrocarbons: an environmental perspective. Microbiolog Rev 45:180-209.

Atlas R, Bartha R.
　1973　Stimulated biodegradation of oil slicks using oleophilic fertilizers. Environ Sci Technol 7:538-541.

Baker J, Little D, Owens E.
　1993　A review of experimental oil spills. In: Proceedings of 1993 Oil Spill Conference, American Petroleum Institute, Washington, DC. p 583-590.

Bartha R.
　1986　Biotechnology of petroleum pollutant biodegradation. Microb Ecol 12:155-172.

George-Ares A, Clark JR, Biddiner GR, Hinman ML.
　1999　Comparison of test methods and early toxicity characterization for five dispersants. Ecotoxicol Environ Safety 42:138-142.

Kelso D, Kendziorek M.
　1991　Alaska's response to the Exxon Valdez oil spill. Environ Sci Technol 25:16-23.

Lee DJ, Portier RJ.
　1999　In situ bioremediation of amines and glycol-contaminated soils using low intervention methods. Remediation 9:117-132.

Lin Q, Mendelssohn IA, Henry CB, Roberts PO, Walsh MM, Overton EB, Portier RJ.
　1999　Effects of bioremediation agents on oil degradation in mineral and sandy salt marsh sediments. Environ Technol 200:825-837.

Lunel T, Rusin J, Halliwell C, Davis L.
　1997　The net environmental benefit of a successful dispersant operation at the Seam Empress incident. In: Proceedings of the 1997 International Oil Spill Conference, American Petroleum Institute, Washington, DC. p 185-194.

NRC [National Research Council].
　1985　Oil in the Sea: Inputs, Fates, and Effects. Washington, DC: National Academy Press.

Overton E, Sharp W, Roberts P.
　1994　Toxicity of petroleum. In: Cockerham L, Shane B, eds. Basic Environmental Toxicology. Boca Raton, FL: CRC Press. p 133-156.

Portier RJ.
  1991 Applications of adapted microorganisms for site remediation of contaminated soil and water. In: Martin AM, ed. Biological Degradation of Wastes. New York: Elsevier. p 247-259.
Portier RJ, Ahmed SI.
  1988 A marine biotechnical approach for coastal and estuarine site remediation and pollution control. Mar Technol Soc J 22:34-42.
Prince RC, Varadaraj R, Fiocco RJ, Lessard RR
  1999 Bioremediation as an oil spill response tool. Environ: 20:891-896.
Pritchard PH, Costa C.
  1991 EPA's Alaska oil spill bioremediation project. Environ Sci Technol 25:372-379.
Smith TG, Portier RJ.
  1997 A risk assessment of chlorinated aliphatics in bioremediated soils. Remediation 7:107-132.
Stewart P, Tedaldi D, Lewis A, Goldman E.
  1993 Biodegradation rates of crude oil in seawater. Water Environ Res 65:845-848.
Walker M, McDonagh M, Albone D, Grigson S, Wilkinson A, Baron G.
  1993 Comparison of observed and predicted changes to oil after spills. In: Proceedings of 1993 Oil Spill Conference. Washington, DC: American Petroleum Institute. p 389-392.
Wolfe MF, Schwartz GBJ, Singaram S, Mielbrecht EE, Tjeerdema RS, Sowby ML.
  1998 Influence of dispersants on the bioavailability of naphthalene from the water-accommodated fraction crude oil to the golden-brown algae, *Isochrysis galbana*. Arch Environ Contam Toxicol 35:274-280.

Constraints on the Use of Bioremediation in Wetlands

Irving A. Mendelssohn

INTRODUCTION

The objective of this brief presentation is to provide an independent assessment of the information gaps and research needs relating to the use of bioremediation in wetlands. The paper is organized as a series of questions and answers, which address the factors that limit or restrict the use of bioremediation in wetlands. It should be noted that within the context of this discussion, I equate bioremediation with biostimulation, the addition of nonmicrobial agents such as fertilizers and soil oxidants to stimulate the degradative capacities of naturally occurring microflora. It is now generally accepted that bioaugmentation, the application of oil-degrading bacteria to a contaminated site, is not useful in the wetland environment because of the abundance of indigenous hydrocarbon degraders in these carbon-rich systems.

IS BIOREMEDIATION APPLICABLE TO WETLANDS?

This basic question has not been adequately addressed. Although bioremediation has been demonstrated in laboratory, greenhouse, and some field trials, extensive field tests in a variety of different wetland types from coastal salt marsh to riparian forested wetlands have not been conducted.

Wetland Biogeochemistry Institute and Department of Oceanography and Coastal Sciences, Louisiana State University, Baton Rouge, LA

Additionally, within any given wetland type, environmental gradients in hydrology, salinity, and soil fertility exist that may constrain the realized efficiency and effectiveness of bioremediation. For example, areas of a wetland that are normally submerged may exhibit lower bioremediation effectiveness than sites experiencing daily inundation due to the more biochemically reduced conditions in the former. As a result, manipulative field experiments and controlled greenhouse studies are needed for a variety of wetland types before general conclusions can be made concerning the applicability of bioremediation in wetlands. Additionally, spills of opportunity should be used, whenever possible, to evaluate bioremediation under real-world conditions. Less emphasis should be placed on microcosm experiments because of the artificiality of this type of system.

WHAT IS THE ROLE OF PHYTOREMEDIATION IN THE BIOREMEDIATION PROCESS?

Because the application of bioremediation within a wetland environment generally occurs in the presence of wetland vegetation, we might ask what role the plants, per se, play in the degradation of the oil. Wetland vegetation could reduce oil concentrations in the soil directly by plant uptake as well as indirectly by maintaining a more suitable soil environment for microbial degradation of the oil. Traditional bioremediation agents such as fertilizers may act not only to directly stimulate microbial activity but also to increase plant growth and thereby indirectly affect plant-mediated controls on oil removal and degradation in the soil. Wetland plants may accelerate oil degradation by oxidizing the substrate by radial oxygen loss from roots and by root carbon leachates that may "kick start" the petroleum degraders into action. In the highly reduced soil of wetlands where oxygen may limit microbial activity, one might question whether bioremediation in the absence of plants will be effective. Thus, research to determine the role of phytoremediation in the bioremediation process is essential. We must answer the question: "Are plants necessary for significant bioremediation in wetlands?"

HOW CAN BIOREMEDIATION BE MAXIMIZED IN WETLANDS?

More research is needed to address applied questions relating to the type and mode of use of various bioremediation agents (biostimulants). Fertilization is arguably the primary bioremediation agent used to treat oil contamination in wetlands. A number of unanswered questions exist regarding nutrient amendments. For example, what nutrient most limits microbial degradation in wetlands—nitrogen or phosphorus? Does this limitation differ among different wetland types and even within a given wetland type? We cannot assume that results from salt marsh bioremediation trials are applicable to oil degradation in freshwater marshes. Can micronutrients become limiting to microbial degradation after macronutrient limitations have been alleviated? What forms of the nutrients should be used—urea, NH_4Cl, NH_4NO_3? Is slow-release fertilizer more effective than soluble forms? Are multiple or split applications more beneficial than single applications? What are the best application methods?

Research concerning the use of nonfertilizer agents such as soil oxidants, surfactants, and dispersants is also required. As previously mentioned, the lack of oxygen in wetlands may be a primary factor constraining maximum oil degradation. The effectiveness of oxidants such as calcium peroxide, nitrate, and manganese oxide, to name a few, in stimulating oil biodegradation in wetlands requires further research. In addition, the use of surfactants and/or dispersants either alone or in combination with fertilizers and soil oxidants to maximize biodegradation needs investigation. The simultaneous use of multiple agents should be considered.

WHAT ARE THE MECHANISMS CONTROLLING EFFECTIVE BIOREMEDIATION?

Basic research that elucidates the primary biotic and abiotic factors controlling oil degradation in wetlands is a first step in developing more effective bioremediation methodologies. How do nutrients, oxygen, temperature, and their interactions limit bioremediation? What is the relationship between the plant and microbial responses relative to oil biodegradation? The effect of the plant rhizosphere, that zone in the soil affected by plant roots, on microbial activity and oil degradation needs considerable more research. The differential effect of various plant species on oil degradation has not been investigated. For example, the question of whether there are plant species-specific differences in capacities to accelerate oil degradation, tolerances to oil, soil oxidative capacity, root architecture and distribution, root exudate release, and rhizosphere development should be addressed. The wetland environment can be complex, with abiotic factors such as salinity, inundation, and pollutants other than oil, which affect the potential for bioremediation. Thus, the effects of multiple and interacting environmental stressors on oil biodegradation requires investigation.

HOW CAN BIOREMEDIATION BE INTEGRATED WITH HABITAT RESTORATION?

A major concern in oil spill response is the integration of oil cleanup and habitat restoration. Can we employ methods that simultaneously accelerate oil degradation and promote habitat restoration? One methodology that has this potential is phytoremediation. The use of marsh plantings both to speed the recovery of the habitat and to accelerate oil degradation by phytoremediation is very appealing. Even if plantings are not needed because the original vegetation survived the spill, the application of fertilizer to increase plant growth rates and vegetative reproduction will accelerate habitat restoration and likely accelerate biodegradation.

WHAT IS THE ROLE OF BIOREMEDIATION IN OIL SPILL RESPONSE?

Immediately after an oil spill, mechanical cleanup is usually the first method employed to remove the bulk oil, if indeed the oil is removed at all. Bioremediation is not a first-response choice because of its inability to degrade large volumes of oil, but rather it is useful as a finishing technique to remove residual oil. Therefore, bioremediation is not the complete answer to oil spill cleanup, but instead is one of a number of methods that may be employed at various stages of the oil spill response to cleanse the environment. Bioremediation, if it is used at all, should be part of an integrated oil spill cleanup response.

WHAT ARE THE NEGATIVE IMPACTS OF BIOREMEDIATION AND HOW CAN THEY BE MITIGATED?

The impact, if any, of the application of bioremediation agents like fertilizers or soil oxidants to the environment must be assessed. This information would allow natural resource trustee agencies and oil spill responders to make informed decisions concerning the potential trade-offs between using a bioremediation agent and allowing the oil to degrade naturally. In addition and if possible, this information can be used to alleviate concerns that the public may have concerning bioremediation. Questions such as the following must be answered: Would large-scale fertilization result in significant coastal eutrophication and harmful algal blooms? Are there ecotoxicological effects of these agents on biota? If negative impacts are identified, how can they be mitigated? The widespread use of bioremediation will likely not be accepted until the potential environmental impacts are adequately addressed.

WHAT IS THE ROLE OF GOVERNMENT AND SOCIETY IN LIMITING THE IMPLEMENTATION OF BIOREMEDIATION?

Federal and state trustee agencies, local government, oil spill responders, oil and gas industry representatives, and the public directly or indirectly determine the widespread acceptance of bioremediation as a cleanup tool for oil spills. Research directed to answering many of the general and specific questions posed above is essential to provide the information required to determine 1) whether bioremediation is effective in oil spill remediation in wetlands, 2) under what conditions it can be maximized and therefore when it should be used, and 3) whether negative environmental impacts can result and what can be done to avoid or reduce these impacts. Once this information is available, and assuming that it is favorable to bioremediation, trustee agencies, interested parties, and the public will likely be much more accepting of this procedure.

HOW CAN WE BETTER INFORM USER GROUPS?

The acceptance of bioremediation by regulatory agencies and the public requires dissemination of the information gained from bioremediation research to both of these groups. Organizations such as the American Petroleum Institute and the National Research Council could be instrumental in this regard.

IS FUNDING FOR RESEARCH IN BIOREMEDIATION ADEQUATE?

To answer the questions above, a concerted research effort in bioremediation is needed. This will require additional sources of funding to those presently available.

SUMMARY

Bioremediation shows potential as an oil spill remediation technique for wetlands. However, considerably more information is needed before this potential can be realized and the effectiveness of bioremediation can be maximized. Funding to support research to address bioremediation in the wetland environment and the dissemination of this information to user groups are essential if we are to see the widespread acceptance of this methodology for oil spill remediation in wetlands.

Restoration

Judith McDowell

INTRODUCTION

We will now shift gears slightly from discussions of bioremediation to more general ecological concerns. Over the past decade, we have seen a marriage of ecology and molecular biology with the hope of better understanding the mechanisms that control the diversity and composition of ecosystems. This combination has resulted in some very insightful ways to look at ecosystems, and the title of this session is restoration. By understanding these mechanisms we will be able to begin to restore damaged ecosystems.

In this section, titled "Restoration," the three speakers will talk about coral reef habitats. Coral reefs are some of the most productive and diverse ecosystems on the earth. They are threatened throughout the world by anthropogenic activities such as shipping traffic, anchor damage, oil spills, and nutrient runoff; and natural processes such as disease, global climate variations, hurricanes, and other storm events. The speakers will introduce you to various aspects of coral reefs and the processes that control coral reefs.

Woods Hole Oceanographic Institution, Woods Hole, MA

Opportunities for Biotechnology for Coral and Reef Restoration

Aileen N. C. Morse

SITUATION AND NEED

Coral reef ecosystems are among the most diverse ecosystems on Earth. Their survival and recurrence over geological time indicate that they possess effective mechanisms of acclimation and adaptation to disturbances. Yet, evidence from recent climatic and episodic events indicates the possibility that these mechanisms are being excessively taxed (Buddemeier and Smith 1999; Done 1999). Complex interactions, as yet hardly understood, between effects resulting from the trend in global warming and those from anthropogenic impacts on near-shore reefs are thought to have led to large-scale changes in community structure, bioerosion, tissue mortality, reduced abundance of corals, and increased incidence of disease (Brown and others 1996; Chadwick-Furman 1996; Jokiel and Coles 1990; Smith and Buddemeier 1992). One of the critical consequences of these disturbances has been to reduce effectively the reproductive potential or capacity of many coral communities. Significant numbers of adult reproductive colonies and young recruits have been partially or totally destroyed (Fisk and Done 1985; McField 1999; Meesters and Bak 1993; Wilkinson and others 1999). This situation is a serious threat to the future stability of coral reefs, given that the integrity and diversity of coral reefs is maintained by processes of sexual reproduction and recruitment

---

Marine Biotechnology Center, Marine Science Institute, University of California, Santa Barbara, CA

of new corals, vegetative spread of individual colonies, and the reestablishment of colony fragments.

Consequently, human intervention as a means to reverse this situation is being considered. Our current understanding of the basic biological and physiological processes and genetics of corals is very limited. It is quite apparent that we simply do not have the technological base to support a concerted effort of reef restoration. If we are to adopt this approach, we must first become better informed about the processes that maintain a healthy reef ecosystem, as well as increase our ability to predict and monitor natural and anthropogenic stresses. These advances will require major technological advances in the areas of coral genetics, coral cultivation, restoration technologies, and molecular physiology. What we do have is a whole range of novel biotechnology approaches recently developed for other areas. Many of these approaches appear to be suitable, after appropriate modification, for direct application to coral reef restoration.

Additionally, at present, there is no concerted global funding effort to facilitate rapid sharing of information, on both developing situations that need to be addressed and the development of novel approaches, particularly modern molecular and genetic approaches, to pinpoint emerging problems and solve them. Most of the currently funded research is being conducted at the individual small-group level. We learn of developments after the fact, through publications. For reefs, which are largely distributed in remote, often underdeveloped areas, this is a particular problem that must be addressed.

OPPORTUNITIES FOR BIOTECHNOLOGY

Genetics

The first, most obvious area providing an opportunity for the development and application of a biotechnological approach is that of the genetics of corals. For coral restoration, the first consideration is to be able to produce a source of coral recruits for out-planting into the reef environment and outgrowth in an aquaculture setting for production of young corals for the aquarium trade. In recent years, there has been mass destruction of reefs, particularly in the Pacific, because of indiscriminate acquisition of fish and corals to supply the huge demand of the worldwide aquarium trade. An important consideration when developing a plan for out-planting of new recruits to the reef is the genetic makeup of existing populations of corals, which is still an open question. Recent ecological evidence predicts that most recruitment occurs locally. Sammarco and Andrews (1988) found that the number of recruited larvae significantly decreased with distance from the source of larvae after a mass-spawning event that occurred on the reefs around an island. Hughes and others (1999), in an experiment covering the length of the Great Barrier Reef, found no recruitment to panels in areas of relatively low density of adults; recruitment occurred in areas of the reef where high densities of adults occurred, suggesting that the majority of larvae were locally retained. In contrast, two different lines of evidence suggest that local recruitment comes from distant supplies of larvae. Studies of gene flow incorporating the genetic composition of local and distant populations of adults indicate that the norm is interbreeding among widely distant populations (Ayre 1990; Ayre and others 1997; Bohonak 1999). Additionally, there is ample evidence for the ability of coral larvae to successfully delay metamorphosis (Morse and others 1996; Morse and others 1988; Richmond 1987). Even after several months in the plankton, larvae retain both their stringency of requirement for an external inducer of metamorphosis and specificity of recognition of the required chemical cue. This ability thus confers fitness for long-range dispersal and recruitment to distant reefs. These results imply that molecular genetic information will be required before selecting sources of brood stock for larval recruits. Additionally, genetic screening of all larvae raised in aquaculture facilities will be required to maintain genetic diversity.

A recent innovation, gene chip technology (Gerhold and others 1999), should prove to be a very powerful tool in this area. This tool can be modified to address several areas pertinent to reef restoration. DNA/DNA hybridization could be used for genotyping (DNA mapping and sequencing) to resolve questions of population structure and diversity. DNA/RNA hybridization (gene expression) analyses could identify prevailing environmental parameters or physiological conditions both on the reef and in an aquaculture facility, enabling detection of altered patterns of gene expression as early-warning indicators of stress.

Coral Cultivation

Restoration of damaged reefs and supply of young corals for the aquarium trade are the two main targets in coral cultivation. The few attempts at stony coral aquaculture (in the Florida Keys) were unsuccessful as viable enterprises. They appear to have failed because of lack of hard scientific data to guide their approach. This situation is very similar to the early history of shellfish aquaculture, particularly that of abalone. The approach was based on anecdotal evidence rather than on scientific fact. Only after funding by the California Sea Grant Program of basic research into the physiology and molecular mechanisms that control reproduction, metamorphosis and grow-out did this industry begin to grow and succeed. There have been prior experimental attempts in Hawaii to reattach coral fragments to small posts, but the lack of success precluded this approach from developing to the commercial stage. Before aquaculture can be considered we will need to develop core technologies for the control of reproduction, larval rearing, metamorphosis, grow-out and outplanting, genetic screening and diversity, and perhaps, in the future, genetic improvement of broodstock. Our purpose will be to produce a multitude of different species in large numbers, under controlled aquaculture conditions. For reef restoration, we will need to produce both complex (branching) and robust (solid) corals; for aquaculture branching, stony corals are the most desirable type.

In many invertebrates, the processes of reproduction and larval metamorphosis are regulated by specific environmental molecular signal molecules (Morse 1991; Morse and Morse 1991b). For many years, these processes have been successfully harnessed for control in finfish and shellfish aquaculture; recent investigations have revealed related pathways in corals. Broadcast spawning, fertilization, planktonic behavior and larval settlement, and metamorphosis all depend on molecular recognition of a factor in the environment. Reproduction and spawning in many molluscs is controlled by prostaglandins (Morse 1984). The activity of this hormone can be mimicked by hydrogen peroxide (Morse and others 1977), a simple chemical adaptable for use in aquaculture of many molluscs (Morse and others 1978). We and our colleagues in Japan recently induced Pacific acroporid corals to spawn using both prostaglandins and hydrogen peroxide. The gametes were viable for fertilization; larvae developed normally and metamorphosed in response to the required chemical cue. These results suggest that techniques for inducing spawning in corals in aquaculture could be based on those developed for molluscs. Until very recently, the processes that coordinate synchronous spawning of multiple colonies of numerous stony coral species has been a mystery. Tarrant and others (1999) found that estrogens appear to act as bioregulators of this process, as well as gametogenesis. Development of inexpensive mimics of the identified estrogens, estrone and estradiol-17 beta, will prove useful for control of these processes in an aquaculture setting.

The signal molecules that control settlement and metamorphosis in marine invertebrates are not as universal as those controlling reproduction are. These are usually species or group specific. Our understanding of these processes in corals is quite developed. The chemical cue required for metamorphosis is a sulfated polysaccharide of the calcified cell walls of crustose red algae, the algae that cement the reef structure together (Morse and Morse 1991a). Representative corals of four major coral families, Acroporidae, Faviidae, Agariciidae, and Poritidae, all require this same chemical cue (Morse 1998; Morse and others 1996). When examined in the light of recent phylogenetic revisions of corals (Chen and others

1995; Romamo and Palumbi 1996; Veron and others 1996), these data reveal a common ancestry for this mechanism that dates back to 240 Ma (Morse and others 1996).

Restoration Technology

To restore corals to damaged reefs, it will be necessary to produce coral larvae in an aquaculture setting, raise them to metamorphic competence, induce larvae to metamorphosis on suitable substrates, and out-plant the young recruits onto the reef. To accomplish this objective, we must develop reliable technologies to predict and control gametogenesis, spawning and larval production, induction of larval settlement and metamorphosis, and successful out-planting. There is a critical need to especially address the first and last of these; we already have the beginning of the technological base for controlling metamorphosis. For restoration purposes, we will be relying on the viability of adult corals brought in from the field as brood stock. In the wake of recent bleaching events, it is clear that the reproductive capacity of corals is negatively affected not only by loss of reproductive mass (Fisk and Done 1985; McField 1999) but also by interference with reproductive physiological functions (Glynn 1996; Morse 1996; Rinkevich 1996; Szmant and Glassman 1990). In addition, reproductive capacity is commonly affected by a number of stresses that include hurricanes, typhoons and storms, elevated water temperatures and increased UV irradiance, elevated nutrient and sediment loads, and disease. There is, therefore, a need for the development of new technologies to predict, assess, and analyze reproductive capacity and factors (intervening environmental factors as well as "stressors") affecting this process. Additionally, we need to update our ability to predict fecundity. To accomplish this, we need to develop simple uniform assays based on the physiological processes involved. Our current methods are unreliable at best.

In spite of recent observations of reduced fecundity, the available number of gametes for potential harvest is enormous compared with our ability to harness this resource. Presently, our needs for laboratory-level gamete capture are adequate (Morse and others 1996), but much more efficient technologies must be developed for aquaculture purposes. Identification of the physiological indicators of timing of release of gametes (and assays for their detection) is critical to efficient capture of gametes that are released on only one night of the year. Methods developed for successful fertilization of gametes and rearing of larvae in the laboratory (Morse and others 1996) must be experimentally modified for aquaculture. In this context, we need to identify environmental factors that limit complete gamete development—a bottleneck in fertilization success. There has been recent success in isolating sperm-attractant molecules from a single montiporid species; three highly unsaturated fatty acids in a particular ratio were identified (Coll and others 1994). Results suggest that particular ratios of related fatty acids might act as sperm attractants in a species-specific manner.

The finding that corals share a common chemosensory mechanism (Morse and others 1996) has made it possible to develop chemoinductive substrates, or "flypapers," with proven efficacy in field tests for successful recruitment of larvae on the reef (Morse and Morse 1996; Morse and others 1994). The purified inducer contains both hydrophobic and ionic moieties, and both properties have guided the development and experimentation of different coupling technologies. The most recent of these uses technologies borrowed from the semiconductor industry. A monolayer of the purified inducer is coupled by a linker to a silanized surface, resulting in a highly potent inductive substrate that is active for long periods in seawater. There is still room for further development of this product; we are working on the flexibility of the substrate material. This flypaper technology is ideal for controlled settlement and metamorphosis of larvae for aquaculture. It is anticipated that these substrates will also provide a means of easily out-planting newly settled recruits onto the reef, which we have repeatedly demonstrated in a research situation (Morse 1998; Raimondi and Morse forthcoming). Additionally, they are potentially useful for resolution of other factors involved in recruitment. Examples include monitoring the availability of larvae for recruitment from the plankton; assessing variation in recruitment under different environmental conditions, one indicator of reef health; and offsetting the collection of corals from reefs for the aquarium, jewelry, and ornamental trades and providing an alternative source of coral for medical purposes such as bone replacement.

The main criteria that will be used to access the outcome of restoration technology will be establishment of a reproductive population of new adult corals. For a given species, this population will comprise a critical number of survivors when corals reach reproductive age. Additionally, maximum long-term growth rates will be a factor—the larger the colony, the greater its potential capacity. Controlled field studies with newly metamorphosed agariciids have allowed us, for example, to determine those criteria for species in this complex. One of the lessons from these studies has been how critical it is to determine in pilot studies what type of habitat confers the greatest growth and survivorship for a particular species (Raimondi and Morse forthcoming). As soon as corals become reproductive, relative measures of their reproductive capacity will be the other criteria. When the reef becomes reasonably well established, we would expect to see an influx of fish, particularly those associated with corals rather than macroalgae.

Particularly in Florida, transplantation of corals from a healthy coral-rich area to one requiring restoration is being considered as an alternate (but not necessarily competing) approach. Attachment of fragments, or even whole corals, to hard substrates with underwater cement is possible. This approach appears to be a possible viable alternative. There are, however, several considerations, particularly when large areas are to be restored. First it means removing large amounts of coral biomass, whether it be composed of multiple fragments or individual adults, because successful fertilization for any one species depends on a critical number of individual colonies in relatively close proximity to one another. This criterion is true for both mass spawning and planulating corals. Judging from the relatively low success of fertilization in the wild on established reefs compared with that obtained by individual crosses in the laboratory, the required number of colonies is high. The other suggestion to save recently dislodged corals, which involves sending teams of volunteers into the field to transport these corals to an aquaculture facility, also has its limitations. Assuming we had determined the culturing conditions, this approach would work only with rather small corals. Larger corals would overwhelm the capability of most systems to effectively remove the nutrient waste produced by any significant number of larger corals. Rather, it would be better to attempt to cement them back on the reef, even with the inevitable loss of some tissue. We recently removed reattached fragments of *Acropora palmata* to a variety of sites to monitor differential survival and growth; there were no survivors after 1 month. So far, there have been no success stories using this approach, but it is worth consideration.

Molecular Physiology

The development of modern technologies for analysis and assessment of the core physiological processes of photosynthesis and symbiosis, reproduction, development, and growth in corals will be required for accurate predictions of stresses that affect corals (Glynn 1993). This type of information is central to successful reef restoration. We need to be able to quickly and efficiently detect and diagnose impending, acute responses of corals to stress that lead to reproductive and recruitment failure, bleaching, and mortality. To date, we have no physiological indicators of impending reduction of the former two processes. Recent studies, however, have identified a number of indicators associated with increased UV radiation and elevated seawater temperatures that result in coral bleaching.

Induced expression of specific heat-shock proteins has been detected in coral tissues before actual bleaching (Black and others 1995; Fang and others 1997; Hayes and King 1995; Jones and others 1998), as has that of UVB-protectant proteins (Gleason and Wellington 1995) and DNA repair enzymes (Lesser 1996, 1997), the production of which are indicative of high levels of UV radiation. For these proteins and the P-450 proteins induced in response to increased pollution and sediment, we currently have only phenotypic screening to screen for their presence. Earlier detection with increased accuracy is now possible by genotypic screening. Technologies are needed to develop molecular diagnostic indicators using, for example, gene chips as suggested above for exquisite detection of altered gene expression.

Deep Reef Corals

Recent investigations by myself and other researchers of populations of corals on deeper reef (70-95 m) slopes in the Caribbean suggest that these relatively disturbance-free communities may provide additional insights into the mechanisms that maintain healthy reefs (Bunkley-Williams and others 1988; Fricke and Meischner 1985; Ghiold and Smith 1990; Goreau and Wells 1967). Additionally, particularly in times of high stress in shallower parts of the reef, they may be sources of larvae for natural replenishment of recruits. In this respect, they may be potentially useful candidates for the study of adaptive mechanisms that facilitate coral colonization outside their normal range. Warner (1997) has suggested that corals may be able to vary the phenotypes of their young to adapt to variable environments. Although such investigations may not be for immediate consideration, they are something to consider for the future.

REFERENCES

Ayre DJ.
   1990   Population subdivision in Australian marine invertebrates: Larval connections versus historical factors. Aust J Ecol 15:403-411.

Ayre DJ, Hughes TP, Standish RJ.
   1997   Genetic differentiation, reproductive mode, and gene flow in the brooding coral *Pocillopora damicornis* along the Great Barrier Reef, Australia. Mar Ecol Prog Ser 159:175-187.

Black NA, Voellmy R, Szmant AM.
   1995   Heat shock protein induction in *Montastrea faveolata* and *Aiptasia pallida* exposed to elevated temperatures. Biol Bull 188:234-240.

Bohonak AJ.
   1999   Dispersal, gene flow, and population structure. Q Rev Biol 74:21-45.

Brown BE, Dunne RP, Chansang H.
   1996   Coral bleaching relative to elevated seawater temperature in the Andaman Sea (Indian Ocean) over the last 50 years. Coral Reefs 15:151-152.

Buddemeier RW, Smith SV.
   1999   Coral adaptation and acclimatization: A most ingenious paradox. Am Zool 39:1-9.
Bunkley-Williams L, Morelock J, Williams EH Jr.
   1988   Lingering effects of the 1987 mass bleaching in Puerto Rican reefs in mid to late 1988. J Aquat Anim Health 3:242-247.
Chadwick-Furman NE.
   1996   Reef coral diversity and global change. Global Change Biol 2:559-568.
Chen CA, Odorico DM, ten Lohuis M, Veron JE, Miller DJ.
   1995   Systematic relationships within the Anthozoa (Cnidaria: Anthozoa) using the 5'-end of the 28S rDNA. Mol Phylogenet Evol 4:175-182.
Coll JC, Bowden BF, Meehan GV, Konig GM, Carroll AR, Tapiolas DM, Alino PM, Wheaten A, De Nyse R.
   1994   Chemical aspects of mass spawning in corals: I. Sperm-attractant molecules in the eggs of the scleractinian coral *Montipora digitata*. Mar Biol 118:177-182.
Done TJ.
   1999   Coral community adaptability to environmental change at the scale of regions, reefs and reef zones. Am Zool 39:66-79.
Fang L-S, Huang S-P, Lin K.
   1997   High temperature induces synthesis of heat-shock proteins and the elevation of intracellular calcium in the coral *Acropora grandis*. Coral Reefs 16:127-131.
Fisk DA, Done TJ.
   1985   Mass bleaching of corals on the Great Barrier Reef. In: Proceedings of the 5th International Coral Reef Symposium 5:149-154.
Fricke H, Meischner D.
   1985   Depth limits of Bermudan (Atlantic Ocean) scleractinian corals: A submersible survey. Mar Biol 88:175-188.
Gerhold D, Rushmore T, Caskey CT.
   1999   DNA chips: Promising toys have become powerful tools. Trends Biochem Sci 24:168-173.
Ghiold J, Smith SH.
   1990   Bleaching and recovery of deep-water, reef-dwelling invertebrates in the Cayman Islands, British West Indies. Carib J Sci 26:52-61.
Gleason DF, Wellington GM.
   1995   Variation in UV-B sensitivity of planula larvae of the coral *Agaricia agaracites* along a depth gradient. Mar Biol 123:693-703.
Glynn PW.
   1993   Coral reef bleaching—Ecological perspectives. Coral Reefs 12:1-17.
Glynn PW.
   1996   Coral reef bleaching: Facts, hypotheses and implications. Global Change Biol 2:459-509.
Goreau T, Wells JW.
   1967   The shallow-water scleractinian of Jamaica: Revised list of species and their vertical distribution and range. Bull Mar Sci 17:442-453.
Hayes RL, King CM.
   1995   Induction of 70-kD heat shock protein in scleractinian corals by elevated temperature: Significance for coral bleaching. Mol Mar Biol Biotech 4:36-42.
Hughes TP, Baird AH, Dinsdale EA, Moltschaniwskyj NA, Pratchett MS, Tanner JE, Willis BE.
   1999   Patterns of recruitment and abundance of corals along the Great Barrier Reef. Nature 397:59-63.

Jokiel Pl, Coles SL.
1990   Response of Hawaiian and other Indo-Pacific reef corals to elevated temperature. Coral Reefs 8:155-162.
Jones RJ, Hoegh-Guldberg O, Larkum AWD, Schreiber U.
1998   Temperature-induced bleaching of corals begins with impairment of the CO_2 fixation mechanism in zoozanthellae. Plant Cell Environ 21:1219-1230.
Lesser MP.
1996   Elevated temperatures and ultraviolet radiation cause oxidative stress and inhibit photosynthesis in symbiotic dinoflagellates. Limnol Ocean 41:271-283.
Lesser MP.
1997   Oxidative stress causes coral bleaching during exposure to elevated temperatures. Coral Reefs 16:187-192.
McField MD.
1999   Coral response during and after mass bleaching in Belize. Bull Mar Sci 64:155-172.
Meesters EH, Bak RPM.
1993   Effects of coral bleaching on tissue regeneration potential and colony survival. Mar Ecol Prog Ser 96:189-198.
Morse ANC.
1991   How do planktonic larvae know where to settle? Am Sci 79:154-167.
Morse ANC.
1996   Effects of natural bleaching event on agariciid larval production, metamorphosis, overall survival and growth. In: Proceedings of the 8th Coral Reef Symposium, Panama.
Morse ANC.
1998   An ancient chemosensory mechanism controls metamorphosis of larvae in divergent coral families. In: Proceedings of the 3rd International Larval Biology Meeting, Melbourne, Australia.
Morse ANC, Iwao K, Baba M, Shimoike K, Hyashibara T, Omori M.
1996   An ancient chemosensory mechanism brings new life to coral reefs. Biol Bull 191:149-154.
Morse ANC, Morse DE.
1996   Flypapers for coral and other planktonic larvae. BioSci 46:254-262.
Morse DE.
1984   Biochemical and genetic engineering for improved production of abalones and other molluscs. Aquaculture 39:263-282.
Morse DE, Duncan H, Hooker N, Morse A.
1977   Hydrogen peroxide induces spawning in molluscs, with activation of prostaglandin-endoperoxide synthetase. Science 196:298-300.
Morse DE, Hooker N, Morse A.
1978   Chemical control of reproduction in bivalve molluscs. III: An inexpensive technique for mariculture of many species. Proc World Maricult Soc 9:543-547.
Morse DE, Hooker N, Morse ANC, Jensen R.
1988   Control of larval metamorphosis and recruitment in sympatric agariciid corals. J Exp Mar Biol Ecol 116:193-217.
Morse DE, Morse ANC.
1991a   Enzymatic characterization of the morphogen recognized by *Agaricia humilis* (scleractinian coral). Biol Bull 181:104-122.
Morse DE, Morse ANC.
1991b   Molecular signals, receptors and genes controlling reproduction, development and growth: Practical applications for improvements in molluscan aquaculture. Bull Inst Zool Acad Sinica Monogr 16:441-454.

Morse DE, Morse ANC, Raimondi PT, Hooker N.
   1994. Morphogen-based chemical flypaper for *Agarica humilis* coral larvae. Biol Bull 186:172-181.

Raimondi PT, Morse ANC. The consequences of complex larval behavior in a coral. Ecology (Forthcoming).

Richmond RH.
   1987   Energetics, competency, and long-distance dispersal of planula larvae of the coral *Pocillopora damicornis*. Mar Biol 93:527-533.

Rinkevich B.
   1996   Do reproduction and regeneration in damaged corals compete for energy allocation? Mar Ecol Prog Ser 146:297-302.

Romamo SL, Palumbi SR.
   1996   Evolution of scleractinian corals inferred from molecular systematics. Science 271:640-642.

Sammarco PW, Andrews JC.
   1988   Localized dispersal and recruitment in great Barrier Reef corals: The Helix experiment. Science 239:1422-1424.

Smith SV, Buddemeier RW.
   1992   Global change and coral reef ecosystems. Ann Rev Ecol Sys 23:89-118.

Szmant AM, Glassman NJ.
   1990   The effects of prolonged bleaching on the tissue biomass and reproduction of the reef coral *Montastrea annularis*. Coral Reefs 8:217-224.

Tarrant AM, Atkinson S, Atkinson MJ.
   1999   Estrone and estradiol-17 beta concentrations in tissue of the scleractinian coral *Montipora verrucosa*. Comp Biochem Physiol 122:85-92.

Veron JEN, Odorico DM, Chen CA, Miller DJ.
   1996   Reassessing evolutionary relationships of scleractinian corals. Coral Reefs 15:1-9.

Warner RR.
   1997   Evolutionary ecology: How to reconcile pelagic dispersal with local adaptation. Coral Reefs 16:115-120.

Wilkinson C, Linden O, Caesar H, Hodgson G, Rubens J, Strong AE.
   1999   Ecological and socioeconomic impacts of 1998 coral mortality in the Indian Ocean: An ENSO impact and a warning of future change? AMBIO 28:111-196.

Coral Epidemiology

Laurie L. Richardson

INTRODUCTION

Coral reefs are exhibiting rapid degradation worldwide. This statement is supported by numerous observations and reports that indicate increasing numbers and types of coral diseases and disease outbreaks (Hayes and Goreau 1998; Richardson 1998); intensifying episodes of coral bleaching with a new trend of bleaching-associated coral mortality (Montgomery and Strong 1994; Williams and Bunkley-Williams 1990); and recently documented, irreversible ecosystem-level shifts from coral dominated to macroalgal-dominated reefs (Done 1992).

Maintaining coral reefs and coral reef health in general is important for numerous compelling reasons. Often cited are economic aspects, which include revenue from tourism as well as physical protection from coastal erosion. Other reasons include the potential of coral reefs as sources for new biomedical and biotechnological substances. Perhaps the most important aspect of preserving coral reefs is based on their intrinsic value as important reservoirs of biodiversity (Reaka-Kudla 1996). This latter factor alone underlines the critical status of coral reefs in that we are now seeing population and community-level shifts in some coral reef systems away from dominance by corals; of imminent worry is the now-realized potential for nonrecovery from such events, termed phase shifts (Done 1992).

Department of Biological Sciences, Florida International University, Miami, FL

The severe and increasing problems in coral reef health are compounded by a lack of information about coral epidemiology. One of the most serious problems is the lack of understanding of coral diseases. Very few coral diseases have been characterized despite the increases in disease incidence and disease-induced coral mortality (Richardson 1998). Confusion and misunderstanding of specific coral diseases and reported disease-like syndromes are prevalent. Yet many uncharacterized "diseases" are currently being monitored on coral reefs, generating a database that is not supported by peer-reviewed research. Also of serious concern is the question of whether the new trend of bleaching-associated coral mortality will continue (Montgomery and Strong 1994). The paucity of understanding of coral (as well as other marine) diseases and mortality events spans local to regional to global scales (Epstein 1998; Harvell and others 1999).

More basic and applied research is needed to understand and thus counteract the observed degradation of coral reefs. This research must address topics ranging from those in which some progress has been made (e.g., characterization of individual coral diseases) to those in which virtually nothing is known (as in the relationships between coral stress, bleaching, disease, and the environment).

STATUS OF THE FIELD

Very little is known about coral disease etiology. Although as many as 15 individual coral diseases have been proposed (most of which are considered to be new or emerging), only five have been characterized beyond the level of observation. Four of these were recently reviewed in an article that focused on coral pathogens and emphasized peer-reviewed results (Richardson 1998). Since this review, a fifth disease has been suggested (Hayes and others 2000). This discussion provides a summarized review of peer-reviewed results on coral diseases and a summary of anecdotal reports of uncharacterized coral diseases.

CORAL DISEASES

The four coral diseases characterized to date are aspergillosis, black band disease, plague, and white band disease. The variety found within these four diseases is fascinating, with a range that spans from a single pathogen (plague and aspergillosis) to pathogenic communities that in turn range from a highly structured microbial consortium (black band disease) to a more loosely organized bacterial community (white band disease). Each of these will be briefly described. For details other than very recent results that are cited in the text, please see the recent review mentioned above (Richardson 1998).

Black Band Disease

Black band disease was the first coral disease to be discovered (in 1973; Figure 1). It consists of a distinctive band that moves across coral colonies while completely destroying coral tissue. The band is composed of a highly structured microbial consortium. Dominant members of the consortium include the cyanobacterium *Phormidium corallyticum*, an oxygenic phototroph; the sulfide-oxidizing bacterium *Beggiatoa*; the sulfate-reducing (and sulfidogenic) bacterium *Desulfovibrio*; and other heterotrophic bacteria. The consortium is directly analogous to laminated microbial mats found in illuminated, sulfide-rich benthic aquatic environments. Within the consortium, the microbial community functions together metabolically to produce and sustain a vertically structured gradient environment (1 mm thick) that is anaerobic and reducing at the base

FIGURE 1. Black band disease on *Diploria strigosa*. The band consists of a microbial consortium that functions synergistically to generate and maintain a highly structured chemical environment that is toxic to coral tissue. The dark color of the band is due to the light-harvesting photosynthetic pigment phycoerythrin present in the cyanobacterial member of the consortium. The entire band community migrates across coral colonies (typically 3-4 mm/day), completely destroying coral tissue and exposing coral skeleton (white area). Colony size = 2 m (height) by 1.5 m (width). Band width = 1 cm.

FIGURE 2. Plague type II on *Dichocoenia stokesii*. This disease, one of the most virulent coral diseases to date, emerged on the reefs of South Florida in the summer of 1995. The disease line starts at the base of colonies and progresses upward to completely destroy coral tissue. It is caused by a potential new species of the bacterium *Sphingomonas*, an aerobic, Gram-negative, flagellated heterotroph. The disease rapidly kills small colonies of corals of 17 species. Colony size = 6 cm (width).

(the result of sulfide production by black band *Desulfovibrio*); oxygen-rich at the surface (due to exposure to oxygen-rich seawater as well as oxygenic photosynthesis by the cyanobacterium); and contains an oxygen/sulfide interface that vertically migrates on a diel basis as a result of varying cyanobacterial photosynthetic activity. It was demonstrated experimentally that the (microbially generated) toxic microenvironment of anoxia and sulfide at the base of the band is lethal to coral tissue.

Plague

Plague, also called white plague, is very different from black band disease (Figure 2). There is no obvious microbial population associated with the migrating, tissue-destroying line; rather, there is a sharp demarcation between freshly exposed coral skeleton and apparently healthy coral tissue. First described in 1977, plague has to date appeared in three forms on the same reefs of the northern Florida Keys. The 1977 form (plague type I) affected six species of coral, killing colonies (at a tissue-destruction rate of 3 mm/day) over periods within 4 months. This form persisted into the 1980s. No pathogen was isolated. In 1995, plague reemerged in a more virulent form (plague type II) that infected 17 species of corals and destroyed coral tissue at the much more rapid rate of up to 2 cm/day. Type II affected small colonies (usually <10 cm) of corals that often exhibited 100% tissue loss within days. The most susceptible coral species, *Dichocoenia stokesii* (Figure 2), exhibited mortality rates of up to 38% of populations within 11 weeks. A combination of microbiologic, microsensor, and genetic techniques was used to identity the pathogen, a bacterium that is most likely a new species of the genus *Sphingomonas* (GenBank accession number AF143861). A possible third form of plague, with identical symptoms to types I and II, emerged on the same reefs in 1999. This form was observed on the largest (>2 m diameter) reef-building corals. Tissue destruction rates were faster than either types I or II. Potential pathogens have been isolated and are in the process of being analyzed using microbiologic, physiological (metabolic), and genetic techniques.

White Band Disease

White band disease was also first noted in 1977. This disease, which targets the reef-building, branching corals of the shallow reef crest, also exhibits a line of tissue destruction that moves across coral colonies. The line of tissue death is variable, giving rise to two forms of white band. White band type I exhibits a sharp demarcation between exposed skeleton and apparently healthy tissue (similar to plague), whereas white band type II may have a zone of bleached (white-appearing) tissue associated with the disease line. The bleached band of type II may, at times, stop advancing, in which case it appears the same as type I. Microbiologic, metabolic, and genetic studies have shown that white band type II is associated with a bacterial community that always contains *Vibrio charcharii*. No consistent pathogen has been found for type I. White band disease is, to date, one of the most destructive of coral diseases and has killed more than 90% of acroporid corals in the Caribbean region. It was shown to have completely restructured one reef in Belize.

Aspergillosis

Aspergillosis of seafans (*Gorgonia* spp.) was the first of a number of apparently new coral diseases to be characterized beyond the descriptive, anecdotal stage. This lesion-producing disease emerged in 1996 in an epizootic that infected 95% of seafans throughout the Caribbean. The disease is caused by a species of terrestrial fungus, *Aspergillus sydowii*. Studies of this disease have made it possible to ascertain the first known successful coral resistance to infection by a pathogen (Kim and others 2000). Research on aspergillosis has also yielded the first information about a coral disease reservoir that spans regional areas approaching a global scale—the presence of spores in Saharan dust that settles in the Caribbean.

Rhodotorulosis

Rhodotorulosis is the second newly emerging disease to be (very recently) suggested (Hayes and others 2000). This disease, also termed rapid wasting syndrome, is hypothesized to be associated with intercellular growth of the pathogenic fungus *Rhodotorula rubra*. Diseased colonies exhibit rapid breakdown of both coral tissue and coral skeleton. As the disease targets reef-building corals, the breakdown of coral skeleton directly degrades the physical integrity of the reef. Laboratory studies determined that coral tissue is broken down by the metabolic activity of the fungus. Although *Rhodotorula rubra* has not yet been shown to be the pathogen associated with this disease, isolation of this species from parrotfish oral secretions supports the hypothesis that rapid wasting syndrome may be initiated by parrotfish bites.

To characterize the five coral diseases described above, it was necessary to use different combinations of multiple techniques in an integrative manner. These techniques included oxygen and sulfide sensitive microelectrodes (to measure the chemical structure and chemical dynamics in black band and plague); microscopy (light, TEM, and confocal laser techniques) and molecular genetics (species-specific probes and sequence analysis) to identify pathogens; and both metabolic (carbon source utilization pattern analysis) and physiological (varied) methods to define the functional roles of the different pathogens. Additional techniques from the fields of geology, population biology (analysis of survey data), and remote sensing were also used in the studies above.

ANECDOTAL OBSERVATIONS

In addition to the diseases above, a number of postulated coral diseases have been described but not characterized in any rigorous detail. These are more appropriately referred to as syndromes. For most, virtually nothing is known beyond visual-based description supported by a photograph. There is currently much confusion with regard to these syndromes. For example, one halo-shaped bleaching pattern is variously reported as ring bleaching, yellow band disease, yellow blotch disease, or the later stages of dark spot disease. In another instance, four very different patterns of red-pigmented microalgae on corals have been termed red band disease.

The pattern of naming and publicizing anomalous coral syndromes has resulted in widespread confusion within the field of coral epidemiology (Epstein 1998; Hayes and Goreau 1998; Richardson 1998). Such anecdotal observations of coral syndromes have gained wide-ranging publicity by way of the internet and the popular literature and have also appeared in non-peer-reviewed segments of peer-reviewed publications. One of these is the *Reef Sites* section of the highly credible, peer-reviewed journal *Coral Reefs*. *Reef Sites*, which normally presents a photograph and a caption, is meant to portray unusual or interesting phenomena on reefs. In some instances, incomplete studies of coral syndromes reported through this venue are cited with a reference to a page number in this prestigious journal. No further work to characterize the potential pathological condition is conducted, and in this way the uncharacterized syndrome is incorporated into citations of characterized diseases.

Both anecdotally described coral syndromes and peer-reviewed characterizations of coral diseases have been incorporated into three different sets of coral disease identification cards that have been produced and widely distributed in the last 3 years. There is contradiction among individually portrayed (noncharacterized) diseases of the same name between the cards. Besides immediate confusion, research funds are being used to support monitoring programs that use different versions of these cards for disease identification. These differences will be problematic in the future in that the resultant data bases will, in part, have no consistent scientific foundation.

AREAS IN WHICH RESEARCH IS NEEDED

There is an obvious need to characterize coral diseases fully. Several additional research areas in the field of coral epidemiology that should also be addressed include determination of the capabilities and mechanisms by which corals are resistant to disease; elucidation of the role of stress in susceptibility to disease; defining the relationships between bleaching and disease; and assessment of potential correlation between environmental factors such as water quality and disease incidence. Also needed is basic research to support a much-needed definition of what, in fact, constitutes a healthy reef (Done 1992).

Very little is known about coral immunology and the potential of corals to resist disease. It is not known whether a coral disease pathogen can invoke an immunological response that targets and destroys a pathogen. It is not known whether immunity can be conferred by exposure to pathogen or is a natural benefit of a healthy, nonstressed coral immune system. A few reports have detailed physical defense mechanisms of corals against pathogen invasion, such as the general property of mucus production (e.g., Hayes and Goreau 1998). The most detailed work in this area to date is the recent work on seafan resistance to aspergillosis, in which it was demonstrated that gorgonians can recognize and encapsulate invading fungal hyphae (Kim and others 1999). It has been difficult to conduct research in the area of coral immunology for two reasons—lack of funding and the rapid emergence of diseases coupled with their often transient and nonrecurring nature.

Another area that demands investigation is the role of stress in disease. It has been postulated (Epstein 1998; Harvell and others 1999; Hayes and Goreau 1998) that the sudden increase in coral diseases may be due to increased susceptibility of corals to infection by pathogens that are normally present, a response to the effects of increasing environmental (both anthropogenic and natural) stress. Virtually no results, however, have been published in this area. A supportive argument can be made that coral bleaching is known to be a common result of different stressors that include increased temperature, ultraviolet radiation, changes in light intensity, and sedimentation (Lesser and others 1995). Thus the global increase in coral bleaching has been attributed to a response to increased stress, in particular increasing sea surface temperatures (Montgomery and Strong 1994). In a parallel manner, it can be hypothesized that stress is a factor that has resulted in the global increase in coral disease. This theory cannot be proved, however, until more is known about coral mechanisms of disease (and pathogen) resistance.

Virtually nothing is known about the relationship between coral bleaching and disease. One group has reported that the bleaching response can be induced by a pathogenic bacterium (Kushmaro and others 1996). On a different level, there are a number of recent observations (C. D. Harvell, Cornell University, 1999, personal communication; L. Richardson, personal observation) of a high incidence of corals that are bleached as well as diseased. It is not known whether such corals are more susceptible to disease as a result of bleaching, more susceptible to bleaching as a result of disease, or more susceptible to both as a result of environmental stress.

Very little is known in general about the relationships between environmental deterioration and disease. Environmental aspects of coral epidemiology are limited, for the most part, to reports of disease incidence and morbidity and mortality rates, which range from infection rates of <1% for black band disease to mortality rates of >95% for aspergillosis and white band disease (Richardson 1998). There is a known positive correlation between relatively high water temperature and both black band disease and plague, and some evidence that black band disease incidence increases with increasing nutrients (Richardson 1998).

Another area in which nothing is known is viral-associated coral diseases. New studies in this area would face an immediate problem in that there are currently no coral tissue cultures that could be used to test marine viruses for pathogenicity, or to inoculate with potential virus-containing samples of diseased tissue. There are no existing genetic probes for pathogenic (to coral) viral nucleic material. New efforts in this research area would have to start by searching for virus particles in infected host (coral) cells or virus-encoded DNA signatures with the help of bioinformation approaches.

Finally, another problem in the study of coral diseases is that of nonculturable pathogens. The application of new molecular techniques such as representational difference analysis to overcome this problem has been recently proposed (Ritchie and others 2000).

DISCUSSION

Progress in the field of coral epidemiology has been relatively rapid in the last few years (Porter 2000), due in large part to a shift in approach from observation-based studies to multidisciplinary studies utilizing different techniques in an integrative manner. Many important questions, however, remain to be answered.

Increasing our understanding of coral diseases at the most basic level is of highest priority. The current situation is one in which misinterpretation of information is prevalent. Widely disseminated anecdotal information does not coincide with what is known and published in the peer-reviewed literature. The new suite of tools and techniques available in the fields of microbiology, and molecular biology in particular, should be used in an integrative manner for coral disease and syndrome characterization (including defining disease processes) with the recognition that different coral diseases have, to date, offered different sets of questions.

Research in coral epidemiology should be coordinated and integrated. This approach is being implemented successfully in the University of South Carolina (Aiken) laboratory of Garriet Smith, who has generated and maintains a current metabolic (Biolog) database for more than 6000 representative cultures of marine bacteria isolated from the water column and both healthy and diseased corals. Genetic information is available for many of these isolates. This work is detailed in Ritchie and others (2000).

Many coral disease outbreaks are sporadic or transient, including those of apparently new diseases. To date, successful regionwide efforts to document (and sample) such outbreaks have been the result of ad hoc collaborations supported by individual laboratories and field stations. A capability for coordinated, multiinvestigator rapid response to disease outbreaks is necessary to define coral epidemiology.

Determination of the relationships between coral disease and degrading environmental conditions (e.g., increasing water temperature, elevated nutrients, changing light regimes, turbidity, dust events), with an emphasis on the effect of multiple stressors (Porter and others 1999), is perhaps the most important and underfunded area of research in coral epidemiology. It may be that coral disease outbreaks can result in irreversible phase shifts, as the first step of this transition is mass coral mortality (Done 1992). An important aspect of this problem is that there are no baseline data defining a healthy reef—thus no scale on which to determine if a reef is stressed (Done 1992).

Environmental aspects of coral reef degradation is a global problem that requires connection between studies carried out at the local, regional, and global scale. Current efforts (such as the Caribbean Coastal Marine Productivity [CARICOMP] network) (Ogden 1997) are few and have not been used to study disease. One of the most promising areas for the future study of coral epidemiology that directly addresses the problem of working at a global scale is the use of satellite remote sensing. Existing sensors such as the LandSat Thematic Mapper and the Advanced Very High Resolution Radiometer (AVHRR) have been used to document reef topographic changes and bleaching incidents, respectively. The new generation of hyperspectral imaging sensors, which contain an entire spectrum of information in each pixel of an image, will provide optically based quantitative data that will allow assessment of reef health status. This new field of endeavor would benefit greatly by including coral disease researchers in remote sensing ground-truth efforts.

ACKNOWLEDGMENTS

I thank my many collaborators who have worked together as an integrative, multidisciplinary team to study coral disease etiology. This research is currently funded by EPA Region 4 grant 14-984298-97-0.

REFERENCES

Done TJ.
    1992   Phase shifts in coral reef communities and their ecological significance. Hydrobiologia 247:121-132.

Epstein PR.
    1998   Marine Ecosystems: Emerging Diseases as Indicators of Change. HEED Global Change Program. p 85.

Harvell CD, Kim K, Burkholder JM, Colwell RR, Epstein PR, Grimes DJ, Hofmann EE, Lipp EK, Osterhaus ADME, Overstreet RM, Porter JW, Smith GW, Vasta GR.
    1999   Emerging marine diseases—Climate links and anthropogenic factors. Science 285:1505-1510.

Hayes RL, Goreau NI.
    1998   The significance of emerging diseases in the tropical coral reef ecosystem. Revista de Biología Tropical 46(Suppl 5):173-185.

Hayes RL, Peters B, Brown A.
    2000   Identification of a fungal infection in reef-building coral attributable to *Rhodotorula rubra*, an emerging pathogen in the tropical marine environment. Hydrobiologia (Forthcoming).

Kim K, Harvell D, Kim PD, Smith GW, Merkel SM.
    2000   Role of secondary chemistry in fungal disease resistance of sea fans (*Gorgonia* spp.). Marine Biol (Forthcoming).

Kushmaro A, Loya Y, Fine M, Rosenberg E.
    1996   Bacterial infection and coral bleaching. Nature 380:396.

Lesser MP, Stochaj WR, Tapley DW, Shick JM.
    1995   Bleaching in coral reef anthozoans: Effects of irradiance, ultraviolet radiation, and temperature on the activities of protective enzymes against active oxygen. Coral Reefs 8:225-232.

Montgomery RS, Strong AE.
    1994   Coral bleaching threatens oceans. EOS 75:145-147.

Ogden J.
    1997   Caribbean Coastal Marine Productivity (CARICOMP): A research and monitoring network of marine laboratories, parks, and reserves. Proceedings of the 8th International Coral Reef Symposium. p I:647-650.

Porter J.
    2000   Diseases in the marine environment. Hydrobiologia (Forthcoming).

Porter JW, Lewis SK, Porter KG.
    1999   The effect of multiple stressors on the Florida Keys coral reef ecosystem: A landscape hypothesis and a physiological test. Limnol Oceanogr 44:941-949.

Reaka-Kudla ML.
    1996   The global biodiversity of coral reefs: A comparison with rain forests. In: Wilson DE, Wilson EO, eds. Biodiversity II: Understanding and Protecting our Natural Resources. Washington, DC: National Academy Press.

Richardson LL.
    1998   Coral diseases: What is really known? Trends Ecol Evolut 13:438-443.

Ritchie KB, Polson SW, Smith GW.
    2000   Microbial disease causation in marine invertebrates: Problems, practices, and future prospects. Hydrobiologia (Forthcoming).

Williams E, Bunkley-Williams L.
    1990   Atoll Res Bull 335, 1:71.

Use of Trace Metals in Marine Bioremediation: A Need for Fundamental Knowledge

François M. M. Morel

Biological processes in seawater are dependent on the supply of a number of chemical elements that serve as nutrients and affected by others that may act as toxicants. Besides the major algal nutrients (nitrogen, phosphorus, and silicon), marine organisms require trace elements—chiefly trace metals such as manganese, iron, cobalt, nickel, copper, and zinc—for growth. We now know, for example, that the iron supply limits phytoplankton productivity of some regions of the oceans. Conversely, the seawater concentration of a metal such as copper is nearly toxic to a number of marine microorganisms and may control, for example, the distribution of important photosynthetic species such as *Prochlorococcus* in the water column. Thus it is clear that *in principle*, we may be able to manipulate the concentration of trace metals in seawater to affect a desired change such as enhancing some natural degradation process, shifting the dominant flora and fauna, or increasing the productivity—all of which may be considered aspects of bioremediation *lato sensu*.

Because the concentration of trace metals like iron and copper are very low in seawater—picomoles to nanomoles per liter at the surface—they are particularly apt to be increased inadvertently or purposefully by humans. Certainly we have increased the concentrations of metals in many bays and estuaries in the same way that we have increased nitrogen or phosphorus concentrations. But we have also increased substantially

---

Princeton Environmental Institute, Department of Geosciences, Princeton University, Princeton, NJ

the supply of many trace elements to the entire North Atlantic and North Pacific Oceans through atmospheric pollution. On that scale of ocean basins, the effect of our nitrogen and phosphorus inputs is almost inconsequential. As a result, we can conceive of fertilizing whole regions of the oceans with iron, whereas fertilizing them with nitrogen and phosphorus would be unfeasible. Thus, the subject of bioremediation using metals has a regional (if not quite global) dimension as well as a local one. I focus here exclusively on the use of metals in bioremediation and do not discuss the issue of remediation of metal pollution. Although it is true that metal pollution may be a problem in some marine systems (perhaps even on an ocean-basin scale), the natural biogeochemical processes that cycle trace metals in the marine environment are sufficiently rapid that stopping the pollution is generally all that is required for remediation. In some cases, of course, dredging of metal-laden sediments may be beneficial or necessary.

Here I examine the question of the use of trace metals in marine bioremediation using examples that span the whole range of spatial and temporal scales. My concern is with the establishment of a knowledge base that would make such bioremediation technically feasible as well as socially and environmentally responsible. I particularly focus on the need for fundamental understanding of marine processes, from the molecular to the ecological scale, and the development of molecular and synoptic tools appropriate to oceanographic research.

Perhaps the most obvious application of bioremediation technology in the marine realm is for the cleanup of oil spills. Stimulating the growth and metabolism of microorganisms that degrade hydrocarbons is certainly feasible on the scale of an oil spill and also one of the few practical options available to us—once prevention and containment have failed. In some instances, additions of nitrogen and phosphorus have been shown to be effective in accelerating the biodegradation of the oil, at least on shore. The relative proportions of these major nutrients to the available organic food source are well known, and it is a relatively simple matter to estimate how much should be added.

Not so for trace elements. Many trace elements are necessary for the growth of oil-degrading bacteria, and some, such as iron or copper, are essential cofactors in the very enzymes that catalyze hydrocarbon degradation. Thus, trace metal additions may well be useful, but we have little quantitative knowledge of how much of any particular trace metal is required. In fact it is possible that the metal content of the oil sometimes results in a concentration of some metal in seawater that is too high. For any trace metal, there is only a relatively narrow range of concentrations that is optimal for the growth of marine microorganisms. Below this range, a trace metal is limiting; above it is toxic.

This dual role of trace metals as essential nutrients and as toxicants poses a major difficulty for designing practical protocols for bioremediation. This difficulty is much amplified by the complicated chemistry of trace metals in seawater. Most bioactive metals (e.g., Fe, Co, Ni, Cu, Zn, and Cd) are known to be complexed by strong organic ligands in surface seawater. Titrating these ligands with relatively small additions of metals can result in very large increases in inorganic metal concentrations and, consequently, in metal toxicity. In some cases, it is the addition of a chelator to decrease the inorganic concentration of some metals, rather than the addition of metals, that may be necessary to stimulate the growth of oil-degrading bacteria. To effect a practical bioremediation of oil spills by manipulating trace metal chemistry in seawater requires that we know not only what metals are required by the target organisms and how much, but also what the concentrations of the metals and of their chelators are in seawater and in the oil.

One of the most common uses of trace metals for bioremediation is found in the control of noxious algal blooms in fresh waters. A widespread technique for controlling the proliferation of unwanted phytoplankton or macrophyte species in lakes and reservoirs is simply to add relatively large doses of copper sulfate. (The technique is of course also used in swimming pools to keep algal growth to a minimum.) Based on empirical data, the copper is added to a level that is toxic to the target species; that species is killed along with much of the microflora and fauna, and it settles to the sediments, taking with it some fraction of the toxic metal. The process is not always efficacious, of course, and sometimes requires repeated addition of copper sulfate. Sometimes it also has unwanted effects on macrofauna such as fish.

It has been suggested that a similar approach could be used to control harmful algal blooms—such as those of toxic dinoflagellate species—in coastal waters (Anderson and Garrison 1997). This may perhaps be a particularly appropriate approach. It is generally thought that the relative concentrations of nutrients, including trace elements, and their availability to the biota may be a key factor in determining which species dominate the assemblage of phytoplankton in a given locale at a given time. This may well be true of noxious or toxic species, and it has been suggested that the apparently increasing frequency of harmful algal blooms may be related to changes in the relative availability of major nutrients and/or trace metals brought about by human activity. Thus, the control of the floral composition of coastal waters by manipulating, or rectifying, trace metal chemistry may indeed be feasible and perhaps be advisable.

Besides the formidable practical problems posed by the much larger areas to be treated and the much more dynamic mixing regime of coastal waters compared with lakes, there are complex chemical and biological issues to contend with, however. A major reason to control noxious algal blooms in coastal waters is to protect the health or edibility of shellfish such as clams or oysters. Thus, a control technology must not only stop the growth of unwanted algal species, it must also preserve the growth of other species that serve as necessary food to the whole ecosystem, including economically important shellfish. The quasicomplete elimination of the aquatic flora that is practiced in small eutrophic lakes cannot be blindly used in coastal ecosystems. An effective metal treatment method requires that we understand thoroughly both the chemistry of trace metals in coastal waters and the role of these metals in the ecology of the phytoplankton. Because of the high concentration of organic compounds in coastal waters, the nature and extent of metal chelation by natural organic complexing agents is even less well understood than it is in open ocean waters. Furthermore, we are just beginning to understand the relationship between trace metals and the physiology of a few species of marine phytoplankton; we are still very far from understanding the ecological role of trace metals in marine systems. Much chemical, biochemical, physiological, and ecological work needs to be done before we can envisage designing and implementing a successful and safe technique for controlling algal blooms by modifying the trace metal chemistry of a coastal area.

The relentless increase in atmospheric carbon dioxide concentration (pCO_2) and its attendant effect on global warming is of concern to an increasing number of people. This concentration is currently 360 parts per million (ppm); it was 270 ppm 100 years ago, before the massive use of fossil fuel necessitated by the industrial revolution. Atmospheric carbon dioxide was even lower, less than 200 ppm, 15,000 years ago, at the height of the last glacial period. What biogeochemical processes made pCO_2 so low during glacial times, and can we take advantage of these processes to check the present increase in pCO_2? According to one major hypothesis, the very low glacial pCO_2 may have been caused by massive fertilization of large regions of the oceans, particularly the Southern Ocean around Antarctica (Price and Morel 1998). This fertilization would have resulted in a more effective uptake of CO_2 by photosynthetic marine organisms and its sequestration in deep oceanic waters upon remineralization of the sinking particulate biomass. Iron, transported to the oceans from the continent (along with other trace metals) by the high winds that characterized glacial times, is the hypothesized fertilizer. Indeed, it has now been demonstrated that addition of iron promotes the growth of phytoplankton in some regions of the world's oceans, including the Southern Ocean.

Some have suggested—sometimes in jest, sometimes in earnest (Chisholm and Morel 1991)—that we could now engineer the largest (premeditated) bioremediation project of all times by fertilizing the Southern Ocean with iron and sequestering the fossil fuel-derived CO_2 into the oceanic abyss. Besides the staggering technical problems posed by the scale of such a project, there are profound questions of scientific uncertainty (as well as troubling moral issues). Our present experience with iron fertilization experiments, which scale from beakers to a few square miles of oceans, gives us insight into the response of the system over only a few days. The added iron practically disappears from the system after such time and is incorporated into the biogeochemical processes that cycle iron and other trace metals in surface seawater. What, in fact, would happen to the ecology of the Southern Ocean if we sustained it with a high level of iron fertilization over several years? We have no idea. On a time scale longer than a few days, the response of the system to a steady addition of a limiting nutrient such as iron would become dependent on complex feedback processes between biology and chemistry (such as the production of new chelating agents whose identity and functions are yet unknown) and be dominated by unpredictable successions of primary producers and consumers. Our present understanding of the role of trace metals in marine ecology is certainly no better on an oceanic scale than it is on the scale of red tides or oil slicks. The possible unforeseen environmental consequences, however, are proportionately much greater.

The three hypothetical examples of bioremediation by trace metals discussed above point to an urgent need for a sound understanding of the relation between the marine biota and its chemical milieu, at the molecular, organismic, ecologic, and oceanographic levels. Our knowledge of the biochemistry and physiology of marine plants and bacteria is very primitive compared with that of their distant terrestrial cousins. For example, we have only recently become aware of the existence of families of photosynthetic marine prokaryotes—*Synechococcus* and *Prochlorococcus*—which are probably the most abundant photosynthetic organisms on earth. Large eukaryotic microalgae such as diatoms and coccolithophores, which are responsible for most of the export of organic carbon to the deep ocean (as well as the bulk of the precipitation of silica and calcium carbonate in the oceans—the process of "reverse weathering"), are evolutionarily very distant from land plants. Their enzymes often bear little homology with those of green plants, and their physiology is poorly understood. There is clearly a need for a sizable research effort on the basic biochemistry and physiology of the microorganisms that are the basis of marine ecosystems. Focusing the efforts of several laboratories on well chosen model experimental organisms (and sequencing their genome) could in a few years lead to a new understanding of how these organisms are adapted to life in the marine environment—an environment characterized chiefly by very low concentrations of nutrients, including trace metals. Such research could also be targeted at developing analytical tools and protocols that would allow oceanographers to probe the physiology of these organisms in the field. New biochemical, genetic or immunological markers may allow us to assess directly what elements may be limiting or toxic to the ambient microflora in a given locale at a given time. With appropriate field research, we then should be able to translate this physiological insight into an understanding of the ecological relationships among major taxa of marine microorganisms.

To provide the areal coverage necessary for an understanding of these physiological and ecological processes at the scale of oceans, the new experimental molecular tools will have to be made inexpensive and easy to use. If we develop the appropriate knowledge base, we should be able to harness new technology such as "gene chips" to great effect in oceanography. However, if we do not begin now to acquire a fundamental understanding of marine processes at a level commensurate with the advances in the basic disciplines (e.g., biology and chemistry), the application to the marine environment of biotechnology any more refined than major nutrient addition (e.g., modulation of trace metal chemistry), is likely to be technically inefficient and perhaps imprudent.

REFERENCES

Anderson DM, Garrison DJ, eds.
   1997   The ecology and oceanography of harmful algal blooms. Limnol Oceanogr 42: 1009-1305.

Chisholm SW, Morel FMM, eds.
   1991   What controls phytoplankton production in nutrient-rich areas of the open sea? Limnol Oceanogr 36:1507-1970.

Price NM, Morel FMM.
   1998   Biological cycling of iron in the ocean. In: Sigel A, Sigel H, eds. Iron Transport and Storage in Microorganisms, Plants, and Animals. Vol 35: Metal Ions in Biological Systems. New York: M. Dekker, Inc.

Microbial Contamination

Jed Fuhrman

I first would like to echo the critical importance of knowing the basic properties of the marine system, as Dr. Morel discussed. Although most of my topic here is contamination—that is, things that we have been adding to the marine environments and problems with those additions—it is going to be absolutely essential that we understand how that marine system works before we could understand how we could stop or solve some of the contamination problems. I briefly discuss that subject, but mostly I cover the subject of microbial contamination of marine environments. I attempt to define it and briefly describe what we are doing now and what we can do in the future to improve it.

When I use the term *contamination,* I mean a release of microorganisms into the environment, usually from released waste products. People use incredible amounts of water. Most of it goes through pumped systems. It becomes mixed with human waste and all kinds of other waste and is dumped back into the water cycle and out into the ocean. A primary concern that most people have is human safety related to disease; but of course, many of us are also very concerned about degrading the habitats in natural systems.

Microbes that cause contamination include bacteria (relatively small [~1 mm linear dimension] cellular prokaryotes) and viruses that are non-cellular and very small (~30 to 200 nm). The viruses do not metabolize without their hosts, which is very important because it denotes that they

---

McCulloch-Crosby Chair of Marine Biology, University of Southern California, Los Angeles, CA

are very different from bacteria, which reproduce and do things on their own. In addition to bacteria and viruses, protists can sometimes be disease organisms, as Dr. Burkholder will discuss.

We do not know very much about degradation of the habitats of natural systems from microbes that we are releasing out there, compared with release of, for example, nutrients or chemicals. There might be some very serious problems in that lack of knowledge. Mostly what we know about is when we put a microbe in the environment and it comes back to us as a possible disease agent, and that is the primary subject of my talk.

EXPOSURE TO CONTAMINANTS

Our main potential exposure is from eating shellfish such as clams and mussels. In filtering huge amounts of water (much better than most filters we make), they filter many microbes that we release into the environment, which are subsequently funneled back to us through these organisms, if eaten. I am not going to talk much about shellfish testing, but it is a very serious concern to many people, mostly handled by federal and state food safety agencies.

We are exposed to microbial contaminants when we have contact with the water, such as during swimming, surfing, and boating. A few people have also talked about aerosols, sometimes seen as a haze on the windshield of a parked car at the beach. It is the oily residue from the sea surface that comes out as sea spray, and there are actually many aerosols that come out of the marine environment from surf and spray. There is some possible concern about aerosols causing microbial contamination, but very little work is being done on that subject.

IMPACT OF CONTAMINANT EXPOSURE

The financial impact of even the perception of microbial contamination in the marine environment is extremely costly because people spend a lot of money to go to the beach, live near the beach, or be involved with recreation somehow associated with the beach. The affected industries include tourism, real estate, and a multitude of support industries, which might be far from the beach itself, such as a place in the Midwest that makes boogie boards, beach towels, or floats. The total economic input that relates to use of the beach is in the billions of dollars.

SOURCES OF CONTAMINATION

The one source of contamination that immediately comes to mind, sewage treatment plant effluent, is actually quite well regulated. People know in general what is coming out from sewage treatment plants, and on the West Coast it tends to be released from some deep pipe offshore (excuse my West Coast slant). In the marine environment on the East Coast, there are typically estuaries, and much of the treated sewage may proceed through an estuary out to sea. Thus, although there is a great deal of close-to-shore exposure, relatively few people go swimming in many of the major estuaries near where the sewage comes out (like New York Harbor).

Although sewage from treatment plants is probably not the most serious exposure problem (but still a concern, especially regarding shellfish), we are actually more concerned these days about runoff and nonpoint sources—rivers or storm drains, the latter especially on the West Coast. These sources are known to be a real problem. A related concern is coastal septic systems, where there is no local sewer system and people's individual septic systems have a connection through ground water to the sea. From these sources, contaminated water is released into the marine environment directly at the shoreline or at an estuary—precisely where people want to go swimming. It is rarely regulated, very poorly understood, and comprises a very large amount of material.

I am aware of only one epidemiology study in which people swimming at the beach in Los Angeles were examined (Haile and others 1999). It provides evidence that swimming in a storm drain compared with 400 m away results in approximately twice the likelihood of getting certain symptoms. Many of the symptoms, which include intestinal ailments, rashes, respiratory problems, and fevers, probably come from viruses and not just bacteria (relevant because of the kind of testing that is done today).

TESTING AND DETECTION OF CONTAMINATION

Water Quality

Testing agencies measure certain kinds of viable bacteria by growing them on Petri dishes: They usually pour 100 mL of water through a filter, put it on a dish with growth medium, and see what grows in 24 hours. Therefore, results require a full day. They usually count bacteria called total coliforms, fecal coliforms, and/or enterococci. Standards are a threshold of allowable bacteria of the various types, and sometimes a ratio is used as well. Such a ratio was recently adopted in California. If the count or ratio exceeds a threshold, then the authorities take some sort of action such as posting the beach (posting warnings) or actually closing the beach to swimming. Different places have different responses to detected contamination.

I will mention the California threshold values and put these numbers in context in a subsequent discussion of how many native marine bacteria are already in the water. The new California thresholds are (1) total coliforms at 10^5 colony forming units (CFU) per liter, (2) fecal coliforms at 4×10^3 CFU per liter, (3) enterococci at 10^3 CFU per liter, or (4) a ratio of total fecal coliforms less than 10 when total coliforms exceed 10^4. Interestingly, these standards were adopted in large part because of the epidemiology study finding that the inclusion of the ratios covered a higher incidence of illness.

This set of standards is one of the few with a real scientific basis. Many older standards were more arbitrary, although they were developed to be as scientific as possible. For example, if an "average" person can become ill from swallowing 1000 bacteria, and will swallow x amount of water swimming, the standard should be such and such. These calculations were difficult or impossible to verify. The epidemiology study verified some of these calculations.

These bacteria are considered indicators because most of the organisms that grow on these plates are not pathogens themselves. Although one would not want to eat the microbes growing on the plates, one could be exposed to many of them and not become ill. They are simply indicators of microorganisms that probably came from feces, to which one would not want to be exposed.

The problem with indicators in the context of bacterial testing is that we know many of these coliforms may be coming from animal sources such as birds. So if you have an estuary that happens to be in a bird corridor, with birds migrating through in large numbers, test results will indicate significant amounts of bacteria in the water. Sometimes those bacteria are not human pathogens and not indicators of human problems; they are an indicator that there have been birds or, in some cases, sea lions or something like that. Although it makes sense to avoid water with significant amounts of animal feces, it is not clear what sorts of illnesses are being avoided or if the same standards should apply. So these indicators are not perfect by any means.

Viruses

I previously mentioned the testing of bacteria at the beach; however, viruses are not routinely monitored at beaches in the United States. I think there are a few places in Europe that are just starting to do it. A classical, "standard" way that people measure viruses is to sample a large quantity of water (~100 gallons) and put it through a charged filter. In seawater the salts shield the charges, so it is necessary to coagulate or otherwise treat the viruses to be caught by the filter. The viruses are extracted into a small volume, the extract is added to a tissue culture of something like monkey kidney cells, and after 1 or 2 weeks (depending on the particular viruses), one looks for plaques where the kidney cells have been killed. Each typically represents one original virus (or a small clump of them).

This process takes some time, but it detects viable pathogenic viruses. These are viruses that will kill those kidney cells; like the bacteria test, this method finds the viable ones. However, there is no established standard for viruses in recreational water. No one knows how many per 100 gallons is safe. More importantly, the long time to obtain results does not help in making a decision to close the beach. If it takes 2 weeks to do a test, it is not helpful to say, "Two weeks ago we should have closed the beach." For this reason, the standard method is not a very practical management tool. However, a very important issue is that when we talk about detection of virus, we have the proverbial "needle in a haystack."

We talked about 10^5 per liter of certain kinds of pathogenic bacteria. One might talk about very small numbers of viruses: A few per liter could be harmful because one ingested virus can cause an illness. With bacteria, it usually takes hundreds or thousands to cause an illness. Thus, when we talk about this, we say that there is a lot of natural background here; and when we talk about detection, we have the needle-in-a-haystack situation. Typically bacteria are 10^9 per liter and viruses are 10^{10} per liter. A photomicrograph taken by epifluorescence microscopy of stained viruses and bacteria (see Fuhrman 1999) graphically demonstrates the abundance, which looks like stars on a very clear night. The big dots are bacteria, and all of the little dots everywhere are viruses. Note that this is from non-contaminated seawater, 10 miles offshore in deep water. These are just the naturally occurring viruses and bacteria.

We do not know what kind of viruses these are, although they seem very important in natural ecological and biogeochemical processes (Fuhrman 1999). We are just starting to identify the kinds of bacteria using new molecular techniques to which Dr. Morel alluded, and we are finding that some of these are not even bacteria. In surface waters, 5% of them are archaea, and in the deep sea, maybe 50% are archaea; and they group phylogenetically with the thermophilic species. The closest cultured relatives to the marine archaea have an optimum temperature of 105°C, and they like hot acid. But these archaea are living in seawater—highly aerobic conditions, cool, normal salt, and so forth. They are not like any other archaea that anyone knows about, which I think is fascinating and important. My main interest is to study these bacteria and these viruses because I believe it is not possible to study the contaminants by themselves in the marine environment without studying the surrounding native organisms. The processes that bacteria use in nature to defend themselves against viruses are probably what is killing off the viruses that we send out to the environment.

Recent Advances in Detection

There have been few changes in the tests for bacteria such as *Escherichia coli*, one kind of fecal coliform. Most of the current tests still require growth for 24 hours, which is still remarkably simple and inexpensive. In California, the sampling is often done by lifeguards; it is not something that requires a huge amount of training, and it is not very expensive or difficult to do. You simply need to have small filter units that are sterile, show someone how to carry out sterile technique, allow the bacteria to grow in incubators, and count which organism is of a certain color on a plate.

There have been significant advances in virus tests. It is now possible to look for their genetic material without waiting for them to grow. The reverse transcriptase polymerase chain reaction (RT PCR) can be used to look for specific pathogens and a variety of viruses in marine environments. The following viruses have been found in marine environments: (1) enteroviruses, which are polio, coxsackie, and echoviruses; (2) hepatitis A; (3) adenovirus; (4) Norwalk virus; and (5) rotavirus. These viruses cause a veritable laundry list of illnesses. The tests are reasonably fast (about a day) but costly (about $1000 per assay if you add up all the costs), and they require highly trained operators. I estimate that there are one dozen people in the United States who could probably perform the test without additional training or experience. Many more could learn, if given a detailed protocol. There are only a few published reports of using this method with marine samples, one being that of Griffin and colleagues (1999), who found a great deal of contamination in the Florida Keys.

One unresolved question is whether the tests measure nonviable viruses. In other words, because the test is looking at their genetic material, it is not known to what extent there is dead RNA lying around. There probably is some, but RNA is rather labile stuff. It does not tend to persist very long, but it could survive. So perhaps to some extent, it is an indicator.

Why then do we not simply rely on these bacterial indicators because, after all, these are just another problem resulting from fecal contamination. Should they not all go together in the water? You release this stuff in the water. Perhaps all the problem components follow each other around like tracers; that is the way an engineer might first think about it.

However, unlike nonbiological tracers that move with the water, viruses and bacteria each have their own physical and biological properties, and they are quite different. Bacteria can repair damage from sunlight. Even though *E. coli* might not thrive in seawater and will probably eventually starve there, if you damage a cell, its repair mechanism could repair it even in seawater.

Bacteria might divide also in a marine environment, depending on the conditions; they could be eaten by protists. Viruses cannot repair their damage because they do not have metabolism. They could adsorb or desorb from particles in ways very different from the bacteria. They could remain pathogenic for months under cold, dark conditions. People have found them in the Northeast, for example, in cold sediments; pathogenic viruses from sewage could last for 1 or more years in a viable condition.

In our laboratory, we have been using RT PCR with samples from Southern California beaches to detect enteroviruses. We have approximately 50 measurements from which we have compared detection of viruses with the standard bacterial test. We find there is very little relationship between them—neither a correlation nor any statistically significant relationship. They do not appear to follow each other. Our results (Noble and Fuhrman 2000) suggest that under some cases, it would be prudent to test for both bacteria and viruses and not just bacterial indicators, as is done today. Virus testing might be considered first at high-use beaches adjacent to storm drains, for example. However, we are not yet ready for that because several aspects require more work; rapid concentration methods of viruses from seawater are necessary. Virus detection methods must be less expensive and more rapid, but still specific and simple. They probably should be made quantitative, which we are starting to do now. If possible, they should also have an indicator of viability.

Consider the current test. Instead of simply taking seawater and running the RT PCR test directly, we start with 20 L of seawater, filter it through a 0.2-µm filter, and then put that filtrate through a concentration unit that filters all the water away but leaves the viruses behind. This process holds particles greater than about 30 nm. The procedure currently takes several hours and ends up with about 50 µl of all the virus-size material from that 20 L of seawater. We run the RT PCR test on a few microliters of that material, followed by gel electrophoresis (about 1 hour). In the end, comparison with positive and negative controls allows us to interpret positive or negative results. Sometimes we find there is some other material in the concentrate that interferes with some of the tests, making them inconclusive (negative with the natural sample, but still negative when the authentic virus is added to the concentrate).

Future Improvements

This whole procedure might be improved by the development of rapid methods to concentrate large volumes of water and maintain high sensitivity. Detection may be sped up with some possible approaches such as probes, called molecular beacons. These probes are internal to the PCR product and have a quencher. When they are bound to their target, they unquench, becoming fluorescent when the target is present. Such probes can be used in the PCR test and detected while the test is being done, allowing for a quick quantitative answer. Taqman is another similar test, and it relies on enzymatic release of a fluorophore from a labeled oligonucleotide probe.

Looking farther into the future, an instant test would be a huge help. Most of you probably have seen that a physician can rub a throat swab on a small plate and test for strep throat in a few seconds. It used to take a day, with sample transport to a remote lab and labor-intensive tests. Imagine if a similar test allows lifeguards to roll an instrument around on the beach, testing for bacteria and maybe even for viruses at different locations. They might sample a few liters, put it in a machine, and get quantitative answers in minutes. It might be possible to develop tests like that for seawater, with sufficient resources invested. The same machine could test drinking water, reservoirs, rivers, among other possibilities, and could also be used to track sources of contamination. It is not hard to imagine it being cost-effective, but the initial investment in development is the difficult part.

Another major need regards standards: We need to know what level is safe and what levels are unsafe; when to close the beach and when not. Our knowledge is greatly lacking in such situations. An obvious possible move in this direction would be more epidemiology studies coupled with measurements of these viruses and bacteria.

Most importantly, we need to understand more about the factors that control microbial contamination. Even if we could be certain about closing a beach, it is more important to be able to learn what is happening and how to mitigate the problems. How do we change the way we are releasing pathogens into the environment—the timing, location, or something—so that we might be able to solve some of these problems? How do we know when it might be safe to go back to the water after it has been closed without having to run these tests, especially if they continue to be expensive?

To answer all of these questions, we need to know much more about what controls these pathogens once they are released into the environment. Regulations now use what might be called engineering-like approaches. Some treat microbes as a conserved component like salinity. In fact, there are new standards, using so-called total maximum daily load (TMDL), that seem to be a great improvement regarding what can be released into the environment. However, in Santa Monica Bay in Los Angeles, for example, they have a single number for the coliforms in the whole bay. The standard calls for a certain amount of coliforms, or less than a certain level, as if the bay is one place; yet you can walk along the beach and get a 10-fold change in coliforms over 100 m in some cases. Obviously, some improvements are still needed. We must be able to look at variability and patchiness, because regulation of these microorganisms requires understanding more about how they are moving and what kind of processes are controlling them.

ADDITIONAL ISSUES

One additional issue regards questions relating to pathogens of marine organisms. Although this is not my main area of expertise, I have heard people talk about epidemics among marine organisms, possibly including endangered species such as marine mammals. Almost certainly these are exacerbated by pollution or some other source that might have stressed the immune system of these animals.

As a separate issue, there is the question of marine organisms as a reservoir for human or terrestrial animal diseases. Here we are talking about viruses or perhaps bacteria that have a terrestrial animal source, which then enters the marine environment, infecting marine organisms and then returning to infect land organisms, possibly including humans. Usually, viruses have one species of host or closely related hosts. Some viruses, however, jump from host to host, such as from pigs to humans. Such jumping to or from marine animals is very poorly understood. Some recognized broad host-range examples include the caliciviruses that include Norwalk-like viruses. Some are reported to have remarkably wide host ranges, even including fish and mammals for certain serotypes. More work is needed before we can say whether the exact same strain jumps that far and might infect humans.

Recent examples of viruses jumping to marine mammals include reports of canine distemper in seals in Europe. If epidemics of marine mammals become more severe, one might imagine a higher probability of infections that jump across species lines. It might become a zoonotic concern to the human population if there is a reservoir in a marine environment that keeps reinfecting something that affects humans. As our population increases and moves closer to the coast, this virus jumping could worsen any problem that might exist.

REFERENCES

Fuhrman JA.
1999 Marine viruses: Biogeochemical and ecological effects. Nature 399:541-548.

Griffin DW, Gibson CJ, Lipp EK, Riley K, Paul JH, Rose JB.
  1999   Detection of viral pathogens by reverse transcriptase PCR and of microbial indicators by standard methods in the canals of the Florida Keys. Appl Environ Microbiol 65:4118-4125.

Haile R, Witte J, Gold M, Cressey R, McGee C, Millikan R, Glasser A, Harawa N, Ervin C, Harmon P, Harper J, Dermand J, Alamillo J, Barret K, Nides M, Wang G.
  1999   The health effects of swimming in ocean water contaminated by storm drain runoff. Epidemiology 10:355-363.

Noble RT, Fuhrman JA.
  2000   Enteroviruses detected by reverse transcription polymerase chain reaction in the coastal waters of Santa Monica Bay, California: Low correlation to bacterial indicator levels. Hydrobiologia (Forthcoming).

Molecular Biology and Biotechnology in Marine Toxicology

Mark E. Hahn and John J. Stegeman

THE PROBLEM

The presence of toxic chemicals in the marine environment has long been recognized as a potential threat to human health and to the health of the oceans. The oceans are the ultimate sink for many chemicals of anthropogenic origin but also are a source of naturally occurring toxins (and pharmacological agents). Certain classes of marine pollutants—especially persistent organic chemicals such as halogenated aromatic hydrocarbons—are globally distributed, occurring even in the most remote areas such as polar regions, the open ocean, and the deep sea (Ballschmiter and others 1997; Muir and others 1988; Stegeman and others 1986). In many coastal areas, the concentrations of chemicals in the environment are extremely high (e.g., Weaver 1984).

Marine pollutants have been classified primarily on the basis of their chemical structure (e.g., polynuclear aromatic hydrocarbons [PAHs]) or original source or use (e.g., petroleum hydrocarbons or pesticides). Increasingly, however, chemicals are being grouped by functional characteristics, i.e., properties related to shared effects or mechanisms of action (e.g., "endocrine disruptors" [Limbird and Taylor 1998; McLachlan 1993; NRC 1999]). The chemical nature and possible human health effects of some marine pollutants have been considered in earlier reports (Ahmed 1991).

Biology Department, Woods Hole Oceanographic Institution, Woods Hole, MA

The field of toxicology is concerned with the interactions of such chemicals with biological systems, from molecules to ecosystems. Toxicology truly is an integrative science (much like oceanography) and is rooted in established, basic disciplines from molecular biology and biochemistry, to physiology, and even to ecology. In many respects, an organism's response to chemicals in the environment is another facet of biochemical adaptation (Hochahka and Somero 1984; Prosser 1986). Much of the current effort in toxicology is aimed at understanding, at the most fundamental level, the mechanisms underlying chemical effects, which should bring the science from a descriptive to a predictive mode.

Research in marine toxicology ultimately seeks to understand, monitor, and predict the effects of contaminants. These three objectives are interrelated (Figure 1), and progress in meeting all three will be aided substantially by the use of molecular and biotechnological approaches. In turn, the understanding gained should help to shape the application of biotechnology in practical approaches to monitoring.

Two features of toxicology, and biological sciences in general, seriously aggravate the challenge to understanding and ultimately predicting the nature and significance of the interaction of pollutants with marine organisms. The first is the **complexity** of the problem. When one considers the number of chemicals that are of potential concern (10^4-10^5) and marine species that exist as possible targets (10^6-10^7 [May 1988; Pimm and others 1995]), it becomes obvious that achieving a satisfactory understanding will not be possible using a species-by-species or chemical-by-chemical approach.

Other sources of complexity include the fact that organisms are usually exposed not to single chemicals but to chemical mixtures, the components of which may interact in unexpected ways. We need to better understand and predict additive, synergistic, or antagonistic interactions between chemicals. We must also consider interactions between chemicals and environmental variables such as salinity, temperature, light, and pressure. Finally, we must consider chemical effects at multiple levels of biological organization.

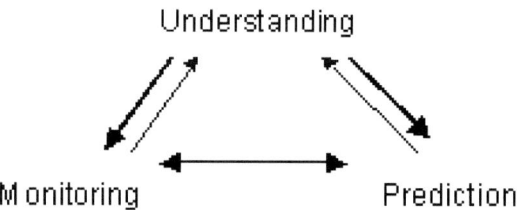

FIGURE 1. The three interrelated objects of marine toxicology research.

The second feature that contributes to the challenge of marine toxicology is **variability**. Marine systems are characterized by geographic and temporal heterogeneity in the concentrations of chemicals and in the structure of communities and ecosystems. In addition, species differences in sensitivity to contaminants and mechanistic variability across taxa complicate the extrapolation of results from one species to another.

CHALLENGES/RESEARCH NEEDS

Progress toward assessing the impacts of contaminants in the marine environment will require a more complete **understanding** of the interactions of these chemicals with biological systems. Such an understanding must occur at the level of molecular mechanisms as well as at higher levels of organization, including populations and ecosystems. It will require basic research to determine the general principles (unity) as well as the detailed differences (diversity) that affect our ability to extrapolate across species or systems.

A mechanistic understanding will stimulate the development of tools for **monitoring** marine organisms for exposure to and effects of chemical contaminants. Such research must go beyond the current approach of monitoring exposure by measurement of chemical residues in biota; the volume of such data has outstripped our ability to interpret it in a biologically meaningful way. It is necessary to move toward biologically based monitoring, by using mechanism-based biomarkers and bioassays (Hahn 2000; Stegeman and others 1992). In addition, monitoring long-term effects of contaminants at higher levels of biological organization will become increasingly important. For example, changes in the population structure or genetic diversity of exposed populations may reveal effects of greater significance for the ecosystem than those measured in individual animals within those populations (Guttman 1994; LeBlanc 1994).

A mechanistic understanding of chemical action and species differences in susceptibility, combined with appropriate monitoring tools, will facilitate the **prediction** of chemical effects (including identification of the most sensitive components of the ecosystem) and the consequences of remediation efforts. Thus, the process of ecological risk assessment will become more accurate, and therefore more useful and less controversial.

Thus, the challenge to marine toxicology, as in other areas of biology, is to extract from the complexity/variability an understanding of the common themes that underlie the responses of the organisms, and conversely to characterize and understand the diversity in organismal responses, and to determine the mechanistic bases for both. Addressing these important questions in marine toxicology can proceed only by basic research, taking full advantage of the advances in molecular biology and biotechnology.

APPROACHES AND EXAMPLES

In this section, we briefly describe some of molecular biotechnological approaches that will be essential to progress in meeting the three objectives outlined above; and in Figure 1, we illustrate their utility with examples drawn from research on halogenated aromatic hydrocarbons (HAHs), a major class of marine pollutants. Although we focus on HAHs, the approaches described will be useful for research on other contaminants as well.

Molecules and Mechanisms

Achieving a mechanistic understanding of chemical effects is a primary goal of research in marine toxicology. Included in this goal is the desire to understand, at a molecular level, the mechanistic basis for species differences in sensitivity to specific contaminants. Much of this information will come from the characterization of the genes—and their protein products—that are involved in toxicity.

An important area of inquiry in marine and aquatic toxicology has emerged from the recognition that many chemicals are toxic by virtue of their ability to interact with intracellular receptors, in some cases mimicking or blocking the effects of natural hormones (Colborn and others 1993; NRC 1999; Schmidt and Bradfield 1996). Toxicity occurs when the resulting changes in gene expression are spatially or temporally inappropriate. The list of such receptors known from studies in laboratory mammals is long, and growing (Table 1). In addition to receptors and other proteins involved in signal transduction, a variety of enzymes modulate the response of organisms to chemical exposure. This variety includes enzymes involved in biotransformation reactions (e.g., the cytochrome P450-dependent monooxygenases) as well as enzymes involved in protective functions (e.g., DNA repair). For each class of proteins, knowledge of their diversity, structure, and function in marine organisms is woefully inadequate relative to the need for such information to understand the impact of chemicals in marine systems (Table 2). For example, although the number of biotransformation enzymes known to exist in humans exceeds 100, fewer than two dozen of these have been studied in any marine organism, and even the most well-studied of these enzymes have been investigated in only a handful of species.

Two types of information will improve our understanding of receptors and enzymes and their role in the toxicity of chemicals to marine organisms. One type is knowledge of the comparative biochemistry and molecular biology of these proteins. Although progress has been made for some of these proteins (Hahn 1998a; Livingstone and Stegeman 1998; Stegeman and Hahn 1994), there is still much to learn (Table 2).

TABLE 1. Soluble Receptors Involved in Xenobiotic Effects

Receptor	No. of forms (genes) per species	Endogenous ligand	Xenobiotic ligands
Aryl hydrocarbon receptor (AHR)	2	Unknown	Dioxins, PCBs[a], PAHs[a]
Estrogen receptor (ER)	2	17-β-estradiol	Organochlorine pesticides; alkylphenols; others
Androgen receptor (AR)	2	Androgens	Organochlorine pesticides; alkylphenols; fungicides
Progesterone receptor (PR)	1	Progestins	Organochlorine pesticides; others
Glucocorticoid receptor (GR)	1	Glucocorticoids	
Constitutive androstane receptor (CAR)	2	Androstanes	Barbiturates; PCBs
Peroxisome-proliferator-activated receptor (PPAR)	3	Fatty acids and metabolites	Fibrates, phthalates
Pregnane X receptor (PXR)	1	Pregnanes, corticosteroids	Organochlorine pesticides; PCBs
Farnesoid X receptor (FXR)	1	Farnesol; bile acids	Unknown
Retinoid receptors (RAR, RXR)	7	Retinoids	Methoprene
Ecdysone receptor (EcR)	1	Ecdysteroids	Hydrazine insecticides
Thyroid hormone receptor (TR)	2	Thyroid hormones	Unknown

[a]PAH, polynuclear aromatic hydrocarbon; PCB, polychlorinated biphenyl.

In investigating toxicologically relevant genes and proteins in marine species, we can expect some surprises in comparison with existing knowledge obtained in terrestrial mammals. Recent findings arising from studies on the mechanism of dioxin toxicity in marine organisms serve as an example. It is well known that the toxicity of chlorinated dioxins and related compounds occurs through activation of a transcription factor known as the aryl hydrocarbon receptor (AHR), and a single AHR gene has been identified in laboratory animals and humans (Schmidt and Bradfield 1996). Recently, a second, novel AHR gene was identified in *Fundulus heteroclitus*, an estuarine fish (Hahn and others 1997; Karchner and others 1999). This second AHR (AHR2) has since been identified in a variety of marine and freshwater fish, and differences in specific sequence motifs and in patterns of expression suggest that the two AHR forms could have distinct functions (Abnet and others 1999; Karchner and others

TABLE 2. Diversity of Genes Involved in Mechanisms of Toxicity: Lack of Information in Marine Species[a]

Gene class (examples)	Number of genes per species (mammals)	Number of genes characterized in marine animals (# species)
Signal transduction (receptors) bHLH-PAS[b]/Ah receptor[b] Nuclear receptors Neurotransmitter receptors	>100	10 (30)
Biotransformation Cytochrome P450s Flavin monooxygenases Transferases	~100	15-20 (30)
Repair/protection DNA repair Oxidoreductases Peroxidases Catalases	~50	0

[a] The numbers listed are order-of-magnitude estimates of the numbers of genes and species for which information is available. These estimates illustrate the diversity of genes involved and the relative lack of information about these genes in marine organisms.
[b] bHLH-PAS, basic-helix-loop-helix-Per-ARNT-sim; Ah, aryl hydrocarbon.

1999). The exact roles of these two AHRs are not yet known but could be related to the sensitivity of fish to dioxin-like compounds. The multiplicity of AHR forms in fish was not predicted from knowledge of mammalian AHRs, illustrating the value of comparative studies in marine organisms, even for molecules that have been well characterized in laboratory species.

The diversity of enzymes and receptors involved in toxicity can be understood—and to a substantial extent predicted—from the evolutionary perspective gained by molecular phylogenetic analysis of genes and gene families. The duplication and diversification of genes within multigene families such as those listed in Tables 1 and 2 is a fascinating area of basic research, which also contributes to a broader understanding of the normal function of those genes and how chemicals might interfere with that function. Recent efforts to determine the evolutionary history of steroid receptors (Escriva and others 1997; Laudet 1997), aryl hydrocarbon receptors (Hahn and others 1997), and cytochrome P450s (Morrison and others 1998; Nelson 1998, 1999) illustrate this approach.

The second type of mechanistic information needed in marine toxicology concerns the in vivo expression of the receptors and enzymes involved in toxicity, and of the genes that are under their control. Until recently, most studies have of necessity focused on measuring expression of single genes, or at most a small number of genes. The emerging field of genomics is changing the way in which we approach such questions. Functional genomic approaches provide a way both to discover the identity of genes whose expression is altered by chemical exposure and to measure those changes rapidly and simultaneously under various conditions (DeRisi and Iyer 1999; Nuwaysir and others 1999). A functional genomic approach to measuring genome-wide changes in gene expression will be a powerful tool for the identification and quantitation of genes whose altered expression leads to toxicity. Although such studies will be restricted initially to model organisms in which extensive genome sequence data exists, this approach will become increasingly applicable and valuable with regard to marine species for which such data are not currently available.

Monitoring: Biomarkers and Bioassays

Analytical chemists have done a superb job of developing exquisitely sensitive methods for detecting contaminants and of applying those methods to generate large databases on contaminant concentrations in a variety of environmental matrices, including marine organisms. Biologists have not been nearly as successful at determining the biological significance of these concentrations.

Monitoring methods based on biological effects and their underlying mechanisms (biomarkers) can complement, and for some applications could replace, the use of analytical chemistry in monitoring the marine environment. The major advantages of such biologic, mechanism-based methods are their toxicological specificity, rapidity, and low cost. Here, "toxicological specificity" refers to the relationship between the assay response and the toxic potential (rather than simply the contaminant concentrations) of the sample being analyzed. McLachlan (1993) called this "functional toxicology." Biological assays include in vivo biomarkers, in vivo bioassays, and in vitro bioassays.

Biomarkers are biochemical, physiologic, or other types of biological changes that indicate the presence or effects of xenobiotic compounds (CBMNRC 1987; Decaprio 1997; Huggett 1992). In addition to the commonly used biomarkers of exposure and effect, which are especially useful in biomonitoring, some biological characteristics can be used as biomarkers of susceptibility (see below). Numerous studies have shown strong relationships between in vivo biomarker responses and exposure to specific classes of marine contaminants (Huggett 1992; Stegeman and others 1992).

Biological and technological advances have facilitated the measurement of biomarkers in marine organisms, and continued improvements are certain. Early measurements of cytochrome P450 1A (CYP1A) induction, which is widely used as a biomarker of exposure to HAH and PAH, relied on enzymatic assays (Payne 1976). The development of monoclonal antibody technology increased the specificity of assays and permitted the analysis of samples obtained from remote regions or those in which enzymatic activity might have been compromised (Stegeman and others 1986; White and others 1994). Analysis of protected species such as whales has been facilitated by recent advances in nondestructive sampling techniques, such as the ability to obtain skin biopsies, and by the identification of vascular endothelium in skin and elsewhere as a prominent site of CYP1A expression (Moore and others 1998). Such nonlethal sampling for biomarkers can be used to establish, on a global scale, the geographic variability in contaminants and their effects in marine mammals or other marine organisms.

In vivo biomarkers rely on natural exposures and responses at the level of the whole organism. Increasingly, mechanism-based bioassays employing whole animals (including transgenic animals), cultured cells, or cellular extracts are being developed and used to detect the presence of contaminants in marine samples (Hahn 2000). Examples include assays that measure receptor-binding, enzyme inhibition, or changes in gene expression. The latter assay can involve native genes such as CYP1A or reporter genes under control of defined enhancer elements that respond to the chemical and receptor of concern. The features of in vivo and in vitro bioassays and their potential for use in monitoring the marine environment have been reviewed recently (Hahn 2000).

Ultimately, a true picture of the effects of contaminants on marine systems will require long-term monitoring of these systems to evaluate changes in community structure, population genetic diversity, and other higher-level features. Evidence for such changes associated with contaminant exposure has been obtained (Guttman 1994), but the consequences of these changes are more difficult to assess. One of the more interesting phenomenon is the emergence of chemical resistance in marine animals after long-term exposure to organic or inorganic contaminants (reviewed by Hahn 1998a; Klerks and Weis 1987; Weis and Weis 1989). The mechanisms underlying such resistance in marine animals remain largely unknown.

Predicting the Impact of Marine Pollutants

Information concerning the identity and concentrations of contaminants in marine biota, along with a detailed understanding of mechanisms of toxicity and the molecular basis for species differences in sensitivity, will allow us to move toward the practical goal of predicting the impact of specific chemicals and combinations of chemicals on marine systems. How might this occur?

One approach to achieving predictive power is through use of biomarkers of chemical effect, such as those described above. By selecting molecular markers that are closely linked to effects of concern, we can minimize the degree of uncertainty in making predictions from biomarker data.

A second approach involves using the proteins involved in causing or modulating toxicity as **biomarkers of susceptibility**. The development of such biomarkers is a natural outgrowth of mechanistic studies. Thus, for example, a receptor such as the AHR can serve as a biomarker of susceptibility to its ligands, in this case the dioxin-like compounds. With respect to human health, biomarkers of susceptibility are often based on interindividual variability in protein function (polymorphisms) and are linked to differences in susceptibility to disease or chemical effects (pharmacogenetics) (Nebert and others 1999; Puga and others 1997). In the context of a marine ecosystem, biomarkers of susceptibility would more likely involve species differences in protein properties that underlie differences in sensitivity. Thus, the sensitivity of marine animals might be inferred by combining information on molecular mechanisms of chemical action with data on the comparative biochemistry of proteins involved in that mechanism.

Molecular and biotechnological methods will be extremely valuable in such efforts. For example, one approach that is being used is the cloning of receptors and enzymes from marine species, followed by in vitro expression and functional analysis of the cloned proteins. Such research, for example, has revealed the catalytic properties of fish P450s, indicating that there may be subtle yet toxicologically important differences between fish and mammals in the active site of CYP1As (Doehmer and others 1999; Oleksiak and others 2000). A similar approach is being used for AHRs. Studies in inbred mice and other species have shown that the expression and properties (e.g., ligand-binding affinity) of the AHR can determine the sensitivity of animals to dioxins; this finding suggests that determining the characteristics of AHRs in marine animals could help predict their sensitivity to dioxin-like compounds (Hahn 1997). Recent studies have shown that the dioxin-binding affinity of the AHR cloned from beluga whales is unusually high, suggesting that beluga, and perhaps cetaceans generally, are among the more sensitive species to effects of these compounds (Jensen and Hahn 1999).

Finally, accurate predictions regarding effects of marine pollutants will require the additional development of mathematical models to describe the behavior of chemicals and the response of organisms to them. Existing models include those predicting the environmental fate of chemicals and the exposure of organisms (Connolly 1991). As our understanding of molecular mechanisms improves, such mechanistic information will be incorporated into more realistic pharmacokinetic and pharmacodynamic models of chemical action (Limbird and Taylor 1998). Demographic models will become increasingly important in translating effects on individual organisms to the level of populations, and eventually, ecosystems (Caswell 1996). The ultimate goal will be to unite these various models to obtain integrated predictions of chemical fate and effects in marine ecosystems.

We face many challenges in our attempts to understand, monitor, and predict the impact of contaminants on marine systems. Molecular and biotechnological methods and approaches will be essential tools in the effort to meet these challenges. Some of the most compelling research needs in marine toxicology are summarized in Table 3 and in the discussion above. Marine and aquatic biologists will continue to look to the biomedical arena for many of the technological advances that will be necessary for this work. In turn, basic and applied research in marine biology, toxicology, and biotechnology will provide tools and reagents that are of great utility in biomedical research (e.g., thermostable polymerases, pharmaceuticals). Such work will also provide a broader, comparative perspective to the study of biochemical adaptation, chemical signaling, and chemical-biological interactions in biological systems.

TABLE 3. Research Needs in the Application of Molecular Biology and Biotechnology to Marine Toxicology

Molecular mechanisms underlying toxic responses and species differences in sensitivity
- Molecular cloning and characterization of genes/proteins involved in toxicity
- Measurement of changes in gene expression on a genome scale (functional genomics)
- Analysis of the molecular evolution of genes, gene families, and pathways

Molecular biomarkers
A. Biomarkers of exposure/effect
- Identification of genes and other biomarkers closely linked to effects of concern
- Validation of target genes for use as biomarkers

B. Biomarkers of susceptibility
- Identification of polymorphisms (pharmacogenetics/pharmacogenomics) and of species differences in gene sequence and protein function linked to differences in chemical sensitivity (e.g., receptors and other signaling proteins)

Mechanism-based bioassays
- In vivo bioassays using transgenic fish
  - Reporter gene-based cell culture bioassays

ACKNOWLEDGMENTS

We thank the Board on Biology, the Ocean Studies Board, and the organizers for the invitation to participate in this workshop. Preparation of this report was supported by NOAA National Sea Grant College Program Office (grant NA46RG0470, Woods Hole Oceanographic Institution [WHOI] Sea Grant project R/B-124, R/B-137 [M.E.H.], and R/P-61 [J.J.S.]), National Institutes of Health grant ES06272 (M.E.H.), US EPA grant R823890 (J.J.S.), and by the Andrew W. Mellon Foundation Endowed Fund for Innovative Research (M.E.H.). Contribution 10114 from the Woods Hole Oceanographic Institution.

REFERENCES

Abnet CC, Tanguay RL, Hahn ME, Heideman W, Peterson RE.
    1999   Two forms of aryl hydrocarbon receptor type 2 in rainbow trout (*Oncorhynchus mykiss*): Evidence for differential expression and enhancer specificity. J Biol Chem 274:15159-15166.

Ahmed FE, ed.
    1991   Seafood Safety. Washington, DC: National Academy Press.

Ballschmiter K, Froescheis O, Jarman WM, Caillet G.
    1997   Contamination of the deep-sea. Mar Poll Bull 34:288-289.

Caswell H.
    1996   Demography meets ecotoxicology: Untangling the population level effects of toxic substances. In: Newman MC, Jagoe CH, eds. Ecotoxicology: A Hierarchical Treatment. Boca Raton, FL: CRC/Lewis Publishers. p 255-292.

Colborn T, Saal FSV, Soto AM.
    1993   Developmental effects of endocrine-disrupting chemicals in wildlife and humans. Environ Health Perspect 101:378-384.

CBMNRC [Committee on Biological Markers of the National Research Council].
    1987   Biological markers in environmental health research. Environ Health Perspect 74:3-9.

Connolly JP.
    1991   Application of a food chain model to polychlorinated biphenyl contamination of the lobster and winter flounder food chains in New Bedford Harbor. Environ Sci Technol 25:760-770.

Decaprio AP.
    1997   Biomarkers: Coming of age for environmental health and risk assessment. Environ Sci Technol 31:1837-1848.

DeRisi JL, Iyer VR.
    1999   Genomics and array technology. Curr Opin Oncol 11:76-9.

Doehmer J, Buters JTM, Luch A, Soballa V, Baird WM, Morrison H, Stegeman JJ, Townsend AJ, Greenlee WF, Glatt HR, Seidel A, Jacob J, Greim H.
    1999   Molecular studies on the toxifying effects by genetically engineered cytochromes P450. Drug Metab Rev 31:423-435.

Escriva H, Safi R, Hanni C, Langlois M-C, Saumitou-Laprade P, Stehelin D, Capron A, Pierce R, Laudet V.
    1997   Ligand binding was acquired during evolution of nuclear receptors. Proc Natl Acad Sci U S A 94:6803-6808.

Guttman SI.
1994 Population genetic structure and ecotoxicology. Environ Health Perspect 102 (Suppl 12):97-100.
Hahn, ME.
2000 Biomarkers and bioassays for detecting contamination of the marine environment. Sci Total Environ (Forthcoming).
Hahn ME.
1998a Mechanisms of innate and acquired resistance to dioxin-like compounds. Rev Toxicol 2:395-443.
Hahn ME.
1998b The aryl hydrocarbon receptor: A comparative perspective. Comp Biochem Physiol 121:C23-C53.
Hahn ME.
1997 The aryl hydrocarbon (Ah) receptor: Biomarker of dioxin susceptibility? In: Proceedings of the Third Finnish Conference of Environmental Sciences, May 9-10, 1997. p 36-39.
Hahn ME, Karchner SI, Shapiro MA, Perera SA.
1997 Molecular evolution of two vertebrate aryl hydrocarbon (dioxin) receptors (AHR1 and AHR2) and the PAS family. Proc Natl Acad Sci U S A 94:13743-13748.
Hochachka PW, Somero GN.
1984 Biochemical Adaptation. Princeton, NJ: Princeton University Press. p 521.
Huggett RJ.
1992 Biomarkers for Chemical Contamination. Boca Raton, FL: CRC Press.
Jensen BA, Hahn ME.
1999 Molecular characterization of a cetacean aryl hydrocarbon receptor, a key protein involved in the toxicity of planar halogenated aromatic hydrocarbons. In: 13th Biennial Conference on the Biology of Marine Mammals, November 28 – December 3, 1999. Society for Marine Mammalogy.
Karchner SI, Powell WH, Hahn ME.
1999 Structural and Functional Characterization of Two Highly Divergent Aryl Hydrocarbon Receptors (AHR1 and AHR2) in the teleost *Fundulus heteroclitus*. Evidence for a novel class of ligand-binding basic helix-loop-helix Per-ARNT-Sim (bHLH-PAS) factors. J Biol Chem 274:33814-33824.
Klerks PL, Weis JS.
1987 Genetic adaptation to heavy metals in aquatic organisms: a review. Environ Poll 45:173-205.
Laudet V.
1997 Evolution of the nuclear receptor superfamily: Early diversification from an ancestral orphan receptor. J Mol Endocrinol 19:207-226.
LeBlanc GA.
1994 Assessing deleterious ecosystem-level effects of environmental pollutants as a means of avoiding evolutionary consequences. Environ Health Perspect 102:266-267.
Limbird LE, Taylor P.
1998 Endocrine disruptors signal the need for receptor models and mechanisms to inform policy. Cell 93:157-163.
Livingstone DR, Stegeman JJ.
1998 Forms and functions of Cytochrome P450. Comp Biochem Physiol 121C:1-412.
May RM.
1988 How many species are there on earth? Science 241:1441-1449.

McLachlan JA.
1993 Functional Toxicology: A new approach to detect biologically active xenobiotics. Environ Health Perspect 101:386-387.
Moore MJ, Miller CA, Weisbrod AV, Shea D, Hamilton PK, Kraus SD, Rowntree VJ, Patenaude N, Stegeman JJ.
1998 Cytochrome P450 1A and chemical contaminants in dermal biopsies of northern and southern right whales. In: Workshop on the Comprehensive Assessment of Right Whales, Cape Town, South Africa, March 19-25, 1998. International Whaling Commission.
Morrison HG, Weil EJ, Karchner SI, Sogin ML, Stegeman JJ.
1998 Molecular cloning of CYP1A from the estuarine fish *Fundulus heteroclitus* and phylogenetic analysis of CYP1A genes: Update with new sequences. Comp Biochem Physiol 121:C231-C240.
Muir DCG, Norstrom RJ, Simon M.
1988 Organochlorine contaminants in arctic food chains: Accumulation of specific polychlorinated biphenyls and chlordane-related compounds. Environ Sci Technol 22:1071-1079.
Nebert DW, Ingelman-Sundberg M, Daly AK.
1999 Genetic epidemiology of environmental toxicity and cancer susceptibility: Human allelic polymorphisms in drug-metabolizing enzyme genes, their functional importance, and nomenclature issues. Drug Metab Rev 31:467-87.
Nelson DR.
1998 Metazoan cytochrome P450 evolution. Comp Biochem Physiol 121:C15-C22.
Nelson DR.
1999 Cytochrome P450 and the individuality of species. Arch Biochem Biophys 369: 1-10.
NRC [National Research Council].
1999 Hormonally Active Agents in the Environment. Washington, DC: National Academy Press.
Nuwaysir EF, Bittner M, Trent J, Barrett JC, Afshari CA.
1999 Microarrays and toxicology: The advent of toxicogenomics. Mol Carcinog 24:153-9.
Oleksiak MF, Wu S, Parker C, Karchner SI, Stegeman JJ, Zeldin DC.
2000 Identification, functional characterization and regulation of a new cytochrome P450 subfamily, the CYP2Ns. J Biol Chem 275:2312-2321.
Payne JF.
1976 Field evaluation of benzopyrene hydroxylase induction as a monitor for marine pollution. Science 191:945-946.
Pimm SL, Russell GJ, Gittleman JL, Brooks TM.
1995 The future of biodiversity. Science 269:347-350.
Prosser CL.
1986 Adaptational Biology. Molecules to Organisms. New York: John Wiley & Sons. 766 p.
Puga A, Nebert DW, McKinnon RA, Menon AG.
1997 Genetic polymorphisms in human drug-metabolizing enzymes: Potential uses of reverse genetics to identify genes of toxicological relevance. Crit Rev Toxicol 27:199-222.
Schmidt JV, Bradfield CA.
1996 Ah receptor signaling pathways. Annu Rev Cell Devel Biol 12:55-89.
Stegeman JJ, Brouwer M, DiGiulio RT, Forlin L, Fowler BM, Sanders BM, Van Veld P.
1992 Molecular responses to environmental contamination: Enzyme and protein systems as indicators of contaminant exposure and effect. In: Huggett RJ, editor. Biomarkers for Chemical Contaminants. Boca Raton, FL: CRC Press. p 237-339.

Stegeman JJ, Hahn ME.
1994 Biochemistry and molecular biology of monooxygenases: Current perspectives on forms, functions, and regulation of cytochrome P450 in aquatic species. In: Malins DC, Ostrander GK, eds. Aquatic Toxicology: Molecular, Biochemical and Cellular Perspectives. Boca Raton, FL: CRC/Lewis. p 87-206.

Stegeman JJ, Kloepper-Sams PJ, Farrington JW.
1986 Monooxygenase induction and chlorobiphenyls in the deep-sea fish *Coryphaenoides armatus*. Science 231:287-1289.

Weaver G.
1984 PCB contamination in and around New Bedford, MA. Environ Sci Technol 18:22A-27A.

Weis JS and Weis P.
1989 Tolerance and stress in a polluted environment. Bioscience 39:89-95.

White RD, Hahn ME, Lockhart WL, Stegeman JJ.
1994 Catalytic and immunochemical characterization of hepatic microsomal cytochromes P450 in beluga whales (*Delphinapterus leucas*). Toxicol Appl Pharmacol 126:45-57.

Critical Needs in Harmful Algal Bloom Research

Joann M. Burkholder

In this presentation, I shall describe areas in which progress is critically needed in the field of harmful algal research. All previous speakers have discussed what they have depicted as "black holes" in basic understanding of the topics they addressed. Harmful algal bloom research is surely another area that could be similarly cast.

"Harmful algae" refers to algae that are undesirable to humans because (1) they produce toxins that impair the health of humans and desirable fish and wildlife; (2) they parasitize desirable organisms in the food web, such as commercially valuable finfish and shellfish; (3) they become too abundant and overgrow desirable habitat for fish such as seagrass meadows, so that the beneficial plants cannot receive enough light to survive; and/or (4) they become too abundant and then, at night, use most or all of the oxygen in the water for their respiration, so that fish and other desirable organisms suffocate or become seriously physiologically stressed. Harmful algae include prokaryotic blue-green algae or cyanobacteria. More recently, the term has been used to include organisms that are not really algae—for example, certain nontoxic animal-like dinoflagellates, which cause fish disease (e.g., *Amyloodinium ocellatum*); and certain toxic animal-like dinoflagellates (e.g., the toxic *Pfiesteria* complex), which do not have their own chloroplasts for photosynthesis but which resemble plant-like dinoflagellates in appearance and certain other general features (Burkholder 1998; Lewitus and others 1999). Here, reluctantly, the cur-

---

Department of Botany, North Carolina State University, Raleigh, NC

rent general misuse of the term *algae* will be followed through inclusion of heterotrophic dinoflagellates under the broad umbrella of harmful algae, although they more correctly should be considered as animal-like protozoans.

Harmful algal blooms have received a great deal of attention, but remarkably little is known about them. This discussion first addresses remote sensing techniques for detecting harmful algae, as requested, and then focuses mostly on critical research needs regarding toxic algal species, as opposed to other types of harmful species that cause oxygen deprivation or other undesirable conditions but do not produce toxins.

ADVANCED TECHNIQUES FOR DETECTING HARMFUL ALGAL BLOOMS

Various remote sensing techniques are available for detecting certain harmful algal blooms, but their value is limited. Remote sensing has helped scientists to track several types of established surface blooms formed by organisms such as certain cyanobacteria, chrysophytes, and dinoflagellates. For example, the toxic dinoflagellate, *Gymnodinium breve*, has been forming blooms in Florida waters for more than 100 years, and it also once bloomed in North Carolina's coastal waters during 1987 (Landsberg and Steidinger 1998; Steidinger and others 1998). In the latter case, it was determined retrospectively that this bloom originated from *G. breve* cells that were transported northward with the Gulf Stream (Steidinger and others 1998). An extremely unusual set of weather conditions allowed small eddies from the Gulf Stream to drift to North Carolina shores basically intact during early autumn. The *G. breve* inoculum increased enough to contaminate shellfish that concentrated them by filter feeding, thus requiring widespread shellfish harvest closures throughout most of the next winter season. That event caused about $26 million of damage to North Carolina; some fishermen never recovered from the losses they sustained.

The analysis tracking *G. breve* northward from Florida (through sea surface temperature patterns) was retrospective. That is, the origin of *G. breve*, once detected in North Carolina waters, was determined belatedly from remote sensing records of temperature patterns from the Gulf Stream. Dense blooms of other toxic algae have also been tracked retrospectively with remote sensing (e.g., Pelaez 1987). However, in general, very little is known about how to prevent blooms, or even to track blooms as they begin to develop. Remote sensing techniques, which would permit design of improved early warning systems for mitigation efforts, are not available to enable detection of initial phases of these blooms. From the perspective of setting early warning systems in motion or mitigating impacts, it would be much more desirable to detect the initiation phase of harmful algal blooms than blooms that are fully developed.

In surveillance, remote sensing techniques are useful in tracking surface water temperatures, salinity, current direction, turbidity, and chlorophyll; however, these techniques have not yet been of use to advance general understanding of factors that control how blooms develop and then dissipate. Remote sensing techniques can sometimes be of value in tracking fully or moderately developed blooms, especially if the harmful species that form them are photosynthetic with plant-like pigments, or if scientists are certain that the bloom distribution closely follows certain temperature patterns or other environmental conditions that can be reliably tracked (e.g., Franks 1995; Johannessen and others 1989).

However, in most estuarine and coastal waters, within practical constraints the currently available remote sensing techniques basically can track only blooms approaching 10 µg or more of chlorophyll/L (Kirk 1994). Thus, such techniques can detect moderate to dense blooms of photosynthetic harmful microalgae and development of undesirable macroalgal growth (e.g., *Enteromorpha* or *Ulva* species in sewage-enriched estuaries). In contrast, some toxic dinoflagellates do not have chlorophyll; and those with chlorophyll sometimes occur in very low cell densities (with chlorophyll a much less than 10 µg/L) that, nonetheless, are sufficient to cause shellfish to become too contaminated with toxins to be safe for human consumption (Falconer 1993a). Remote sensing would not be adequate to track these organisms; nor, in many situations, can the available techniques distinguish between harmful species and other co-occurring benign species with similar pigments. Remote sensing techniques, then, are of use primarily to track environmental conditions that may be associated with harmful algae.

CRITICAL RESEARCH NEEDS

The four most critical research needs are research-quality cultures, life cycles, toxin identification and detection, and detection of toxic strains. Brief discussions are provided for each.

Research-Quality Cultures

From most of the species that have been rigorously tested—from diverse groups including toxic cyanobacteria, chrysophytes, diatoms, and dinoflagellates—it has been established that within a toxic species there is actually a range in toxicity (e.g., Anderson 1991; Bates and others 1998; Burkholder and Glasgow 1997; Edvardsen and Paasche 1998; Gentien and

Arzul 1990; Gorham and Carmichael 1988; Skulberg and others 1993; Sperr and Doucette 1996). Some strains within a population of a "toxic species" can be benign, that is, unable to produce toxin or producing negligible/ undetectable toxin. Moreover, many toxic strains lose toxin-producing capability when maintained in culture for more than several weeks to months, apparently as an artifact of the (highly artificial) culture conditions (e.g., Bates and others 1998; Edvardsen and Paasche 1998; EPA 1999). As they shift from toxic strains to strains that show no detectable ability to produce toxin further, these strains also undergo fundamental changes in physiological and behavioral characteristics.

The danger inherent in misuse of noninducible or "permanently nontoxic" cultures (cultures in which toxicity can no longer be induced; EPA 1999), ostensibly to gain insights about toxic strains of harmful algae, is illustrated by the following example. The toxic dinoflagellate *Pfiesteria piscicida* is a complex animal-like organism, as mentioned (Burkholder and Glasgow 1997). Its response to nutrient enrichment depends on its previous history of feeding, rather than following a typical growth curve with concentration of nutrient added. *Pfiesteria* is stimulated to produce toxin by the presence of live fish (hence the name of the first known species, *piscicida*, meaning "fish killer" as reported by Steidinger and others 1996; also see Burkholder and Glasgow 1997, Burkholder and others 1992, Fairey and others 1999). However, this organism extends retention of kleptochloroplasts from algal prey, is attracted to light in plant-like behavior, and shows minimal attraction to fish after it becomes noninducible (i.e., unable to stress or kill fish in repeated bioassays) over several months in culture with live fish. These are profound changes.

Biohazard III containment systems are required to culture toxic *Pfiesteria* with fish to protect laboratory workers from its aerosolized neurotoxins that fish-killing cultures apparently emit (Glasgow and others 1995). To avoid use of (expensive) biohazard III facilities, *Pfiesteria* can be cultured with algal prey in a temporarily nontoxic mode. However, if toxic strains are cultured for several weeks on algal prey without live fish, most lose their ability to produce toxin. Some strains have been tested repeatedly over 4 years in various culture conditions, and their loss of toxin-producing capability appears to be permanently noninducible, given the present state of knowledge about *Pfiesteria* species (WHOI 2000).

Recently, millions of dollars have been directed to federal agencies to address the toxic *Pfiesteria* issue (Epstein 1998). In the past 2 years, however, much funding has been spent for research on noninducible strains, fed for months to years on algal prey without live fish, which have been supplied to the scientific community at large by a federally funded, national phytoplankton center in the northeastern United States that specializes in growth of plant-like algae. In repeated tests by independent laboratories, these noninducible strains of *Pfiesteria* have proven to be incapable of causing fish stress, disease, or death. Their increased reliance on chloroplasts retained from algal prey makes them more plant-like than toxic strains (Lewitus and others 1999), and they respond to indirect stimulation by nutrient enrichment (mediated through algal prey) more strongly than toxic strains under certain conditions—data with important ramifications if used as planned by state and federal agencies as a basis for setting levels of nutrient reductions to discourage *Pfiesteria* growth (e.g., State of Maryland 1998).

The poultry industry in the Chesapeake Bay region has opposed recent state and federal efforts to reduce nutrient loading from this industry to the Bay. Toxic *Pfiesteria* has been shown to be stimulated directly by nutrient enrichment, thus indicating an enrichment "connection" (Burkholder and Glasgow 1997). However, use of noninducible *Pfiesteria* strains would *bias* studies in favor of finding stronger indirect stimulation of *Pfiesteria* by inorganic nutrients, mediated through the abundance of algal prey that respond directly to the nutrients. This information would not be well received by the poultry industry or others whom government agencies have attempted to move toward strengthened (and costly) nutrient controls, in part by invoking the *Pfiesteria*/nutrient linkage. Thus, the compromised validity of research findings about the behavior, ecology, and physiology of *toxic Pfiesteria* that were erroneously based on use of permanently nontoxic strains could be compounded by serious socioeconomic ramifications that would be avoidable if the importance of toxic versus permanently benign strains is considered in the research design. Similarly, other findings about the behavior and ecology of toxic *Pfiesteria*, based on strains that are incapable of producing toxin, would be questionable.

The case of *Pfiesteria* provides but one illustration of a serious problem that is affecting the general field of harmful algal research. The previously mentioned phytoplankton culture center is endorsed by a consortium of federal agencies and commercially supplies cultures of many toxic algal species to the scientific community at large. The cultures commonly are contaminated with other algal species (e.g., Oldach and others 2000). Some of the cultures have been maintained for many years and are not checked or are infrequently checked to determine whether the strains are still capable of toxin production. Yet, the scientific community relies heavily on these cultures for use in research to further understanding about the behavior, physiology, and ecology of toxic algae.

To avoid compromise of the validity of scientific insights about the physiology, behavior, and ecology of toxic algae, cultures of all "toxic algae" commercially provided for use by the general scientific community should be tested frequently to verify toxic activity. Laboratories that are relied on to commercially supply cultures for research on toxic algae should be required to obtain fresh field isolates as often as is necessary to ensure that fundamental traits (especially the ability of the strain to sustain toxic activity) are maintained in the cultures. Alternatively, arrangements should be made with a laboratory that specializes in production of clonal cultures with demonstrated toxicity to check cultures of the national phytoplankton center frequently to ensure that they still are capable of toxic activity.

The issue of quality control/quality assurance is of major importance in this context. Laboratories that state expertise (e.g., in grant proposals and letters of intent) in providing clonal cultures, techniques, toxin, or other products/services in harmful algal research should be required to provide supporting evidence of expertise in the form of peer-reviewed international science publications on the specific subject, or demonstration of cross-confirmation of their data by a second laboratory with such expertise, or both. Although this stipulation of quality control/quality assurance may seem obvious, it unfortunately is not being required by many federal grant programs in harmful algal bloom research (e.g., indicated in correspondence from ECOHAB-funded scientists expressing concern about the culture quality issue, to the NOAA Coastal Ocean Program, July 1999). Toward the goal of advancing knowledge about toxic strains of algal species—the strains that are germane from the perspective of public concern—assured availability of research-quality cultures is of critical importance and it needs to be more rigorously addressed by the consortium of federal agencies involved.

Life Cycles

Many of the problems inherent in developing techniques to track harmful algae, beyond established blooms of certain photosynthetic species, are grounded in lack of basic information about their biology. The reality is that scientists do not understand the various forms or stages that many of these species can assume (Burkholder 1998; Table 1). The life cycles are poorly characterized and poorly understood. If the range of forms is not known for many of these taxa, then it is difficult to identify or track them, especially with certain techniques in light microscopy that remain in wide use.

For example, of the approximate total of ca. 60 toxic dinoflagellate species (Burkholder 1998), most of the life cycles are incompletely known (Table 2). Many dinoflagellates have animal-like traits (Schnepf and Elbrächter 1992). About half of the described species are heterotrophs, without their own chloroplasts, and about half are plantlike with chloroplasts. Moreover, many of the plantlike species have well-developed

TABLE 1. Toxic Dinoflagellates with Characterized or Partially Characterized Sexual Life Cycles[a]

Species	Gametes	Sexual induction	Sexual cyst
*Alexandrium catenella	Heterothallic, isogamous	Low N or P	+
*Alexandrium tamarense	Heterothallic?; isogamous, anisogamous	Low N or P, increased temperature	+
Amphidinium carterae	Homothallic, isogamous	Increased salinity, photoperiod?	+
Amphidinium operculatum	Homothallic?; isogamous (anisogamous)	Increased salinity	Not known
Coolia monotis	Homothallic, isogamous	Increased nutrients?	+
Dinophysis acuta	Anisogamous?	Factors not known	Not known
Gambierdiscus toxicus	Isogamous	Factors not known	Not known
*Gonyaulax monilata	Heterothallic?, isogamous	Low N	+
Gymnodinium breve	Homothallic or heterothallic, isogamous	Factors not known	+
Gymnodinium catenatum	Heterothallic, isogamous	Low N and P	+
Pfiesteria piscicida	Heterothallic, anisogamous	Dying fish, organic P	+
Pfiesteria shumwayae sp. nov.	Heterothallic?, anisogamous	Dying fish, organic P	+
Prorocentrum lima	Homothallic?, isogamous	Organic P	+
Prorocentrum micans	Isogamous, anisogamous	High nutrients?, temperature shock	+
Pyrodinium bahamense	Heterothallic?, isogamous	Factors not known	Hypnozygote

[a] Sexual reproduction has been suspected or confirmed in 15 of the ca. 60 toxic dinoflagellates known. Asterisks (*) indicate the 3 species for which the life cycle has been reported as completely characterized. Species are indicated as homothallic (with gametes of the same mating type) or heterothallic (having gametes from different [+,-] types). Question marks (?) indicate observations that are suspected but have not been verified. Flagellated gametes area of similar (isogamous) or different (anisogamous) size; in the case of *Prorocentrum micans*, gamete protoplasts unite through a mucilaginous tube (conjugation). N indicates nitrogen; P indicates phosphorus. Hypnozygotes are resting-stage products of sexual reproduction. Adapted from Walker and Steidinger 1979; Nakajima and others 1981; Steidinger 1983; Faust 1992, 1993; MacKenzie 1992; Anderson and others 1998 and references therein.

TABLE 2. Harmful Estuarine and Coastal Marine Microalgae That Have Been Linked to Anthropogenic Nutrient Enrichment

Harmful species	Link to cultural eutrophication
Chattonella antiqua	Bloomed under cumulative high loading of poorly treated sewage and other wastes, coinciding with human population growth (Japan; fish kills, toxic).
Chrysochromulina polylepis	Toxic outbreaks followed change in nutrient supply ratios from cumulative increased nutrient loading (Europe; fish kills, toxic).
Gymnodinium mikimotoi	Bloomed under cumulative high loading of poorly treated sewage and other wastes, coinciding with human population growth (Japan, as *G. nagasakiense*; fish kills, PSP).
Gonyaulax polygramma	Bloomed under cumulative high loading of poorly treated sewage and other wastes, coinciding with human population growth (Japan; fish kills from oxygen depletion).
Nocriluca scintillans	Bloomed under cumulative high loading of poorly treated sewage and other wastes, coinciding with human population growth (Japan; fish kills from oxygen depletion).
Nodularia spumigena	Blooms followed change in nutrient supply ratios from cumulative increased nutrient loading by sewage, agricultural wastes (Baltic Sea; estuary in Australia).
Toxic *Pfiesteria* complex (*P. piscicida, P. shumwayae* sp. nov.)	Most kills [with highest cell densities] have occurred in P- and N-enriched estuaries (e.g., near phosphate mining, sewage inputs, or animal waste operations); between kill events, can prey upon flagellated algae that are stimulated by inorganic nutrients; bloomed 1 wk after a major swine effluent lagoon rupture (with extremely high phosphorus and ammonium) into an estuary, in a location where high abundance of these dinoflagellates previously had not been documented; highly correlated with phytoplankton biomass in other eutrophic estuaries (mid-Atlantic and southeastern United States; fish kills, epizootics).
Phaeocystis spp.	Bloomed following cumulative high loading of poorly treated sewage (Europe; fish—*Phaeocystis pouchetii*); blooms were correlated with altered N/P ratios from cumulative increased nutrient loading (*P. pouchetii*); bloomed 1 week after a major swine effluent lagoon rupture into a eutrophic estuary (*Phaeocystis globosa*, along with *Pfiesteria piscicida*; southeastern United States).

TABLE 2. Continued

Harmful species	Link to cultural eutrophication
Protocentrium minimum	Bloomed under cumulative high loading of poorly treated sewage and other wastes, coinciding with human population growth (Japan; fish kills, toxic); blooms coincide with cumulative high loading of N from sewage, agricultural runoff, atmospheric loading, etc. (southeastern United States).
Prymnesium parvum	Toxic outbreaks usually have occurred under eutrophic conditions (fish kills).
Toxic *Pseudo-nitzschia* complex species	Have occurred with sewage and other wastes (Canada; ASP); consistent seasonal blooms in the Mississippi and Atchafalyu River plume areas, associated with hypereutrophic conditions and in Prince Edward Island, Canada, following anthropogenic nutrient loading and drought).†

†Note that many of the known harmful estuarine and marine microalgae and heterotrophic or animal-like dinoflagellates also have been shown to be stimulated by N and/or P enrichment in culture, which is expected since they are photosynthetic. Also note that blooms of the toxic *Pseudo-nitzschia* complex have not been associated with cultural eutrophication in the northwestern United States. Adapted from Burkholder 1998 and references therein.

heterotrophic capabilities (Hansen 1998). Some of these species have proven difficult or not yet possible to grow or maintain in culture, probably because unknown organic substances needed for growth are not available in laboratory conditions. Similarly, some toxic chrysophytes are known to have amoeboid stages that, thus far, have not been successfully maintained in culture (e.g., Estep and McIntyre 1989).

As additional examples, other harmful (but not toxic) dinoflagellates include certain species that parasitize finfish, shellfish, zooplankton, and benign algae (Cachon and Cachon 1987). The life cycles of most parasitic dinoflagellates are completely unknown (Cachon and Cachon 1987; Pfiester and Popovský 1978). Many of the species remain to be described from one to two stages that can be recognized to date. Successful culture requires the prey, which complicates cloning procedures, especially if the prey are larger organisms (e.g., fish) with a suite of associated contaminating microorganisms and if fresh prey must continually be supplied to the parasites, so that prey sterilization becomes impractical. The environmental requirements of many of the free-living life cycle stages of most parasitic species are poorly understood. Additional information is needed for development of suitable culture media.

Thus, for some harmful algae, culture media needed for successful growth and maintenance of the various life stages have not yet been developed. Obviously, such limitations translate into severe restrictions in what scientists are presently able to learn about these organisms under experimental laboratory conditions.

A related problem merits mention. Historically, research on toxic algae was conducted using defined culture media that had been developed to grow photosynthetic species, that is, strict autotrophs or auxotrophs. Unfortunately, such media narrowly constrained these species so that their heterotrophic behavior was missed (e.g., Jacobson and Anderson 1996). The type and abundance of available food sources have been shown to strongly influence the stages or forms that are present in the (few) toxic dinoflagellates that have been rigorously examined with an array of potential food sources (e.g., Burkholder and Glasgow 1997). Thus, in restricting the nutritional mode of these organisms, various stages in their life cycles may not have manifested (Popovský and Pfiester 1990).

Characterization of the life cycles of many harmful algal species is a critical research need. This information is of fundamental importance to enable scientists to determine ecological controls on bloom dynamics (e.g., nutrient enrichment; Table 3) and to design improved techniques for tracking both planktonic and benthic stages of these organisms. Techniques that may be especially useful in addressing this critical need include the following:

• Low-pressure (high-vacuum) scanning electron microscopy enables live samples to be viewed so that in-progress transformations can be observed at high resolution.

• Additional gene probes and other markers discern various stages (e.g., green fluorescent probes). Fluorescently labeled molecular probes are useful for discerning species of interest among many other species and assorted "debris" in field samples (e.g., Scholin 1998a).

• Gene-specific and other *toxin* probes are critically needed for many harmful species (also see below). Such probes would be of value in verifying toxic stages within the life cycles of harmful species that, in turn, would enable determination of the range of stages that are most important to detect and track.

• The various probes require greater speed in application, more automation, and more amenability to field use than current techniques.

A note of caution is warranted: Although molecular probes used for species identifications are often considered as species-specific, the possibility remains that the probes will cross-react with closely related species that have not yet been tested (Gallagher 1998). Thus, use of probe technology for establishing species identifications should be cross-confirmed with scanning electron microscopy of morphological traits whenever possible, at least on a "spot-check" basis.

TABLE 3. Status of General Knowledge about Toxic Algae[a]

Feature	Toxic dinoflagellates	Toxic chrysophytes	Toxic blue-greens (cyanobacteria)
Life cycles	~ 60 species, many incomplete or unknown; complex life cycle in 2 spp. (*Pfiesteria*); but heterotrophic food sources (aide in detecting complex life cycles) not yet tested for most species	~ 15 species; 1 of 3 *Pseudo-nitzschia* spp. known; 9 chrysophyte spp. poorly or incompletely known (complex life cycles, amoeboid stages in some spp.); 2 of 8 raphidophyte spp. known	~ 40 species, with simplistic (prokaryotic) life cycles known or suspected in most
Toxins	> 50 known; 21 saxitoxins + derivatives characterized; 9 brevetoxins characterized; 4 ciguatera-toxins well characterized (many poorly known), 4 DSP toxins characterized; ≥ 2 *Pfiesteria* toxins partially characterized	In diatoms, domoic acid + isomers characterized; in chrysophytes, several galactolipids, octadecapentaenoic acid, prymnesins, and other uncharacterized toxins; in raphidophytes, hemolytic substances (e.g., polyunsaturated fatty acids), superoxide radicals mostly poorly characterized; other uncharacterized toxins	> 70 known including > 50 microcystins, 3 anatoxins, 16 saxitoxins, 5 opsins, 2 cylindrosperm opsins, 1 nodularin, 1 ciguatera-like toxin
Purified toxins [availability]	Major saxitoxins, brevetoxins commercially available; okadaic acid + several other DSP toxins and derivatives available	Domoic acid commercially available	Major microcystins, anatoxins, saxitoxins, and nodularin commercially available
Toxin assays (for rapid field or laboratory use)	Saxitoxins, brevetoxins as groups (major forms); more limited ciguatoxin assays	Domoic acid commercially available	Major microcystins saxitoxins as groups; nodularin

[a]Diatoms are considered here within division Chrysophyta as class Bacillariophyceae. Adapted from Falconer 1993a; Hallegraeff and others 1995; Burkholder 1998 and references therein; and Chorus and Bartram 1999 and references therein.

Toxin Identification and Detection

As Dr. Fuhrman (this volume) pointed out regarding harmful viruses, there can be substantial economic and social ramifications in harmful algal issues. For example, conventional potable water treatment procedures generally do not remove toxins from freshwater blue-green algae (Falconer 1993a). Moreover, there are no antidotes for the toxins of most harmful algal species, some of which are among the most potent biotoxins known. Before scientists can understand the chronic impacts of these toxins on aquatic organisms and human health, there is a critical need to identify more of these toxins chemically and to develop improved methods for their detection (Hallegraeff and others 1995). Toxin identification and the development of rapid, field-amenable, reliable detection procedures are extremely critical needs in the field of harmful algal bloom research.

A brief outline of the present state of knowledge about the toxins of several major groups of harmful algae and dinoflagellates is warranted (Table 3). Cyanobacteria produce hepatotoxins, and neurotoxins (Falconer 1993b). They are mostly alkaloids, small peptides and one organophosphate. Scientists understand the mode of action of some of the toxins, especially microcystin toxins, for example, which attack the liver and induce hemorrhaging as well as death of liver cells (Carmichael 1994). More than 40 microcystins are known, and there are many other types of cyanobacterial toxins. Standards or purified material are available for a limited number of these toxins—that is, purified toxin material that can be used to conduct research, develop probes for the toxins, and track them in organisms and the environment (Carmichael 1994; Falconer 1993a). Thus, for some of the major microcystins, nodularin, and certain others, these rapid, reliable toxin assays have been developed; but such assays are not available for many cyanobacterial toxins.

In dinoflagellate toxin research, saxitoxins and brevetoxins are well characterized chemically, and their modes of action are fairly well understood (Falconer 1993a). About 20 saxitoxins and their derivatives are known from species found in many parts of the world (e.g., New England coastal waters, the tropics, and Alaska [Falconer 1993a; Hallegraeff 1993]). Standards and assays are available to detect the more common among these toxins. The chronic and sublethal impacts of saxitoxins on humans and aquatic organisms nonetheless remain poorly known (e.g., Burkholder 1998, Landsberg 1996). Saxitoxin-producing dinoflagellates (e.g., the toxic *Alexandrium* complex) affect many geographic regions, including the northeast and west coasts (including Alaska) of the United States (Hällegraeff 1993; Scholin 1998b). Brevetoxins mostly are produced by *Gymnodinium breve* from Florida waters as previously mentioned.

There are approximately nine known brevetoxins, with standards and assays available for several (Cembella and others 1995). Ciguatoxins (also called ciguateratoxins) are more problematic, in part because many are known, including both lipid-soluble and water-soluble substances (Bagnis 1993). Purified toxin standards are available for very few of these toxins; and few assays are available similarly for rapid, reliable detection of certain ciguatoxins in field water samples and animal tissues (Bagnis 1993; Lewis 1995).

Ciguatoxins provide an illustration of the pervasive lack of scientific knowledge about harmful algal toxins, including their detection and range of impacts. Ciguatera (health condition caused by poisoning from ciguatoxins) is the primary source of human poisoning from finfish consumption worldwide (Bagnis 1993; Russel and Egen 1991). Ciguatoxins are neurotoxins that can have "crossover" effects in sensitizing the immune system (reviewed in Burkholder 1998). Their chronic and sublethal impacts have been known to affect people for 10 or more years after acute symptoms subside. However, there are no programs for tracking the chronic/sublethal impacts of these toxins—or chronic impacts from nearly all other algal toxins—on human health (e.g., Burkholder 1998; Hokama and Miyahara 1986; Russel and Egen 1991). Thus, assays generally are not available for use in early warning systems to help people determine where and when finfish contaminated with ciguatoxins will occur. Instead, in many economically depressed tropical regions of the world where this problem is widespread, health officials distribute pamphlets that advise against consuming large barracuda or groupers. Despite important advances within the past two decades, progress in the design of rapid, reliable, field-applicable techniques in ciguatoxin detection has been limited. The resulting limitations in knowledge present serious obstacles for development of improved management strategies to more proactively protect people from toxin exposure.

Thus, even among some of the best known of dinoflagellate toxins, there remain serious limitations in availability of purified toxin standards and rapid, reliable field assays for detection. Diarrhetic shellfish poisoning (DSP) is caused by dinoflagellates that produce another group of toxins, one of which is called okadaic acid (Aune and Yndestad 1993). Chronic/sublethal exposure to this toxin can promote malignant tumors and immune system suppression in mammals, including human tissues (Haystead and others 1989; Hokama and Miyahara 1991). DSP commonly occurs in northern Europe from human consumption of toxin-contaminated mussels and other shellfish (Hallegraeff 1993; Hallegraeff and others 1995). The acute impact is diarrhea, but the potential for malignant tumors and other potential chronic impacts is indicated by laboratory data. Unfortunately, there have been no medical or epidemiological studies to determine whether such impacts are occurring in human populations chronically exposed to okadaic acid. In fact, such studies have not been conducted for most algal toxin exposures. A second general problem in assessing range of impacts is that, as mentioned, many algae toxins remain only partially characterized, without available purified standards or detection assays (Fairey and others 1999; Falconer 1993a; Hallegraeff and others 1995).

In addition to the critical need to fully characterize more of the toxins from harmful algae, assays are also greatly needed to enable rapid, routine, reliable detection of these toxins in potable water supplies, natural waters, seafood, and aquaculture facilities. A probe for domoic acid, for example, together with species-specific molecular probes to verify the presence of the toxic algae that produce it, was valuable in relating the recent sea lion disease and die-off in California to diatoms in the toxic *Pseudo-nitzschia* complex (Scholin and others 2000). Although assays for the better known toxins—saxitoxins, brevetoxins, okadaic acid and certain other (DSP) toxins, and certain ciguateratoxins—are commercially available (e.g., Hallegraeff and others 1995), they are limited in ability to reliably detect more than a few of the toxins that are targeted. In other words, the commercially available assays for saxitoxins cannot be used to detect all of the major saxitoxins; those available for DSP toxins fail to detect all of the major DSP toxins; and so forth. Other limitations in quantification or specificity have led state and federal agencies involved in seafood safety issues to forego relying on these assays in favor of the traditionally used (but less sensitive) mouse bioassay (Cembella and others 1995). Concerted research to develop improved rapid, reliable assays for detecting algal toxins is critically needed to advance understanding about impacts from harmful algae on human health and natural resources.

Scientists must also strengthen insights about how these toxins can be harnessed more effectively for beneficial medicinal use. For example, *Pfiesteria* toxins have been reported to cause profound learning disabilities manifested as short-term memory loss in mostly reversible impacts (Glasgow and others 1995; Grattan and others 1998), with indication of involvement of the hippocampus as a target region of the central nervous system. They also have been experimentally shown to cause severe learning disabilities in small mammal studies (rats; Levin and others 1999). These toxins could be of value in research to advance understanding of human memory function, but they are, as yet, incompletely characterized (Fairey and others 1999). Their chemical structures (identity) must be obtained before modes of action in affecting human health can be determined with certainty.

Detection of Toxic Strains

Although various techniques are in hand (e.g., immunoassays, enzyme assays, neuroreceptor binding assays, cytotoxicity assays) to detect toxic strains in laboratory cultures, the available technology is much more limited for use in detecting and tracking toxic strains within mixed field populations. Molecular probes, where available, can detect the species but cannot discern toxic status. Assays for rapid detection of certain toxins have been developed, as mentioned—although with limitations (Table 3). These assays enable detection of the presence of toxin in waters, shellfish tissues, and other materials that are directly sampled (Hallegraeff and others 1995). However, in practical use the available assays do not make it possible to discern between toxic and nontoxic strains in natural phytoplankton or benthic algal samples.

If the toxic members within a population could be tracked in field conditions, insights about environmental controls on the toxic strains, which are of primary interest in natural resource and health issues, could be strengthened. Scientists could better understand how a range of organisms across aquatic food webs is exposed. Impacts of exposure could be more accurately tracked through seafood (Falconer 1993a) and through food webs (e.g., Shumway 1995), and improved diagnostics for human health effects could be designed. Thus, additional techniques to enable detection and tracking of toxic strains in field populations are needed.

PRESENT STATUS OF CONTROL AND PREDICTION

Basic research on strategies to control harmful algal blooms has been limited, but such research remains in primitive status (Boesch and others 1997). Many harmful algae are detected in a reactive rather than proactive response mode, in part because of the sporadic occurrence of these species, and scientists are currently faced with the problem of attempting to develop control strategies for species that they basically know very little about—at least, for many harmful algae. Each of the three standard types of control strategies—physical or mechanical, chemical, and biological—have been considered for controlling harmful algal blooms. In general, they have not worked well except in limited situations.

For example, physical mixing of the water to disrupt the density-based stratification has been used to minimize noxious freshwater blue-green algal (cyanobacterial) growth in ponds and small lakes, and sometimes in aquaculture facilities (Ross and Lembi 1999). As another example, death of striped bass was averted in coastal aquaculture facilities in initial stages of toxic *Pfiesteria* outbreaks by rapidly replacing the water (and thereby removing the toxic dinoflagellates) in brackish ponds where the fish were being grown. In most larger lakes, estuaries, and coastal marine waters, such techniques generally have not been feasible. There are a few notable exceptions; for example, salmon pens simply have been moved away from waters containing harmful diatom or toxic flagellated chrysophyte blooms (Boesch and others 1997).

In the realm of chemical controls, poisons such as copper sulfate have been applied for many years in small lakes, and in some aquaculture facilities under limited circumstances, to control noxious freshwater cyanobacteria (Ross and Lembi 1999). Bleach pellets have been added to the bottom-accumulated sediments in drained coastal ponds and aquaria to eliminate cysts of *Pfiesteria*, before addition of cultured striped bass and flounder. In most larger lakes, estuaries, and coastal marine waters, such physical and chemical techniques generally have not been feasible. As an additional problem, although many kinds of chemical poisons can be used to kill harmful algae, they are not species-specific. Thus, their use is usually impractical because many beneficial, cooccurring species would also be destroyed (Taylor and Pollingher 1987).

Another type of chemical control, reduction of nutrient pollution (although more difficult to accomplish for socioeconomic reasons) has proven highly successful in minimizing the growth of certain harmful algal species. The best success incidents have been documented for toxic freshwater cyanobacteria in which phosphorus reduction to lakes (varying in size from small lakes to Lake Erie) has significantly reduced growth of the undesirable algae (Wetzel 1983). In small ponds, the same effect sometimes has been accomplished by adding nitrogen fertilizer to increase the N:P ratio and encourage growth of desirable green algae (with high nitrogen optima) rather than noxious cyanobacteria with high phosphorus requirements (Ross and Lembi 1999). In certain poorly flushed estuaries and marine coastal embayments, some harmful algal species have been linked to stimulation by nutrient pollution (Table 2). As a result, long-term strategies targeting nutrient reductions are under consideration or, in the case of the Chesapeake Bay and *Pfiesteria*, are being imposed (State of Maryland 1998).

Biocontrol may be the most promising of control strategies, but it remains the least understood. Certain cyanophages are under consideration for control of noxious cyanobacteria under limited conditions (e.g., certain cyanobacteria strains in small ponds; Lembi and others 1988). A virus with potential for reducing blooms of brown-tide organisms has been discovered (Milligan and Cosper 1994). A dinoflagellate from the Pacific Northwest was found to attack a certain harmful dinoflagellate species under culture conditions, but the latter species occurs in the southeastern United States (Taylor 1987). The beneficial versus detrimental effects of attempts to introduce the dinoflagellate from the Northwest to the Southeast, and the feasibility of being able to achieve success in controlling the southeastern species through this effort, are unknown.

Recently there has been much discussion about the use of clay additions to control harmful algal species, a practice that, although primitive, has been used with success in certain aquaculture operations in the Orient (Anderson 1997; Pérez and Martin 1999). Clay particles adhere to the mucilage of certain algal species (e.g., montmorillonite clays adsorb cyanobacteria; reviewed in Burkholder 1992). The organisms coflocculate or settle out with the clay, and some species of algae can be killed in this way. Major concerns in using such control techniques are impairment of shellfish feeding, clogging of zooplankton apparati, and clogging of finfish gills. Aside from attempts to minimize these potentially serious impacts, the degree of success of such an approach depends on the type of clay used and the algal species in question. Some harmful algal species such as cyanobacteria are susceptible to reduction by clay additions, whereas others are not. Noxious freshwater species of *Anabaena* (including toxic *Anabaena circinalis* and *A. flos-aquae*) were highly susceptible to sedimentation and subsequent death with certain clays such as montmorillonite (Avnimelech and others 1982; Burkholder and Cuker 1991). However, addition of another clay common to the area, kaolinite, proved beneficial to these organisms, which grew well after settling out (Burkholder and Cuker 1991). The algae apparently benefited, as well, from high phosphorus supplies that were adsorbed to the clay. In contrast, Yu and others (1995) reported that dinoflagellate species *Prorocentrum minimum* (sometimes toxic to shellfish; see review by Landsberg 1996) and nontoxic *Noctiluca scintillans* were more adversely affected by coagulation with kaolinite than with montmorillonite.

Some "naked" (unarmored) dinoflagellate species lacking protection from thick cell wall-like coverings appear to be especially vulnerable to cofloccuation with clays (e.g., various *Gymnodinium* and *Gyrodinium* species), and if they adsorb directly to the clay particles, they are destroyed (Burkholder 1992). However, dinoflagellates have a remarkable ability to rapidly form temporary cysts (Taylor 1987). The naked species rapidly excrete copious mucilage that surrounds the cells. At the same time, the organisms take up nutrients that were adsorbed to the clay particles. In this way, some species actually appear to derive benefit from the clay (Burkholder 1992). If there is a small area on the outer thick, mucilaginous cell covering that is left uncovered by the clay, then once the water column is cleared, the dinoflagellate protoplast emerges through that area. Armored dinoflagellates use their outer covering of cellulose plates with membranes as a protective barrier, and "molt" the outer covering with adsorbed clay once the water column is clear. Thus, many algae can survive clay-loading events and apparently can actually benefit from nutrients adsorbed to the clay particles. Such mechanisms for survival and benefit are logical especially in estuaries and certain turbid coastal environments. Clay applications can work under limited conditions in certain circumstances but may cause other problems inasmuch as they can promote potentially serious detriment to desirable aquatic life such as sensitive filter-feeding shellfish species (e.g., Howell and Shelton 1970).

In discouraging but realistic writing, Boesch and others (1997) made the following statement in a publication that was cosponsored by NOAA and the National Fish and Wildlife Foundation: "It is premature to conclude whether control strategies are feasible, applicable or advisable because there is insufficient information to judge effectiveness and weigh benefits against costs." Thus, control is in primitive status within the realm of harmful algal blooms.

Prediction obviously is in similar status, given scientists' fundamental lack of knowledge about many harmful algal species and the current limited technologies for recognizing and tracking toxic strains. Some known conducive environmental conditions can be tracked; however, many of the factors that influence these blooms, especially the nutritional ecology of the algal species and controlling biological interactions, are not known. The exception of progress in prediction is cyanobacteria blooms in freshwater ecosystems (Wetzel 1983). Nonetheless, the degree of toxicity of these bloom formers is difficult to predict. Toxicity can be highly variable from strain to strain within the same cyanobacteria bloom (Gorham and Carmichael 1988), also true of toxic prymnesiophytes, toxic diatoms, and some toxic dinoflagellates as mentioned. Moreover, the environmental signals that trigger toxicity are unknown for nearly all toxic species.

SUMMARY

The most important basic challenges in the field of harmful algal bloom research are, first, to fundamentally ensure that research-quality cultures are available for use by the scientific community at large, which presently is *not* the case, especially for toxic algal species and despite the intent of federal consortium agencies engaged in supporting such effort. Second, the life cycles of many of these species need to be characterized so that they can be recognized and tracked in various stages. Armed with that information, scientists will be able to determine much more about their occurrence and behaviors. Third, more of the toxins and toxin derivatives from these organisms need to be identified so that assays can be developed to track the toxins through the food web, to improve diagnostics for human health exposure and animal exposure, and to determine modes of action. Effective medical treatment will remain beyond reach until the modes of action (i.e., the metabolic pathways in which these toxins function) are understood—information that can be obtained with certainty only after the toxins are identified. These data would also enable scientists engaged in medical research to more effectively harness the potential beneficial uses of the toxins. Fourth, improved techniques are needed for detecting toxic strains among field populations of these organisms, which typically have benign (non-toxin-producing) as well as toxic strains that are physiologically and behaviorally distinct.

Until scientists know much more about the life cycles and toxins of these organisms, harmful algae will remain in the realm of the enigmatic, difficult for the public to understand. Without the fundamental information that can be provided only through the critically needed research that was identified here, the panic that is fostered by lack of understanding (e.g., "economic halo effects" as described in Epstein 1998) will continue to occur—along with all of the hardship that such panic creates wherever people depend heavily on the affected freshwater, estuarine, and marine resources for economic sustainability.

REFERENCES

Anderson DM.
   1991   Toxin variability in *Alexandrium*. In: Granél E, Sundstrom B, Edler L, Anderson DM, eds. Toxic Marine Phytoplankton. New York: Elseview. p 41-51.

Anderson DM.
   1997   Turning back the harmful red tide. Nature 388:513-514.

Anderson DM, Cembella AD, Hallegraeff GM, eds.
   1998   Physiological Ecology of Harmful Algal Blooms. NATO ASI Series G. Ecological Sciences. Vol 41. New York: Springer-Verlag.

Aune T, Yndestad M.
   1993   Diarrhetic shellfish poisoning. In: Falconer IR, ed., Algal Toxins in Seafood and Drinking Water. New York: Academic Press. p 87-104.

Avnimelech Y, Troeger BW, Reed LW.
   1982   Mutual flocculation of algae and clay: Evidence and implications. Science 216:63-65.

Bagnis R.
   1993   Ciguatera fish poisoning. In: Falconer IR, ed. Algal Toxins in Seafood and Drinking Water. New York: Academic Press. p 105-115.

Bates SS, Garrison DL, Horner RA.
   1998   Bloom dynamics and ecophysiology of domoic acid-producing *Pseudo-nitzschia* species. In: Anderson DM, Cembella AD, Hallegraeff AD, eds. Physiological Ecology of Harmful Algal Blooms. NATO ASI Series G. Ecological Sciences. Vol. 41. New York: Springer-Verlag. p 267-292.

Boesch DF, Anderson DM, Horner RA, Shumway SE, Tester PA, Whitledge TE.
   1997   Harmful Algal Blooms in Coastal Waters: Options for Prevention, Control and Mitigation. Washington, DC: US Department of Commerce and US Department of the Interior.

Burkholder JM.
   1992   Phytoplankton and episodic suspended sediment loading: Phosphate partitioning and mechanisms for survival. Limnol Oceanogr 37:974-988.

Burkholder JM.
1998 Implications of harmful marine microalgae and heterotrophic dinoflagellates in management of sustainable marine fisheries. Ecol Appl 8:S37-S62.
Burkholder JM, Cuker BE.
1991 Response of periphyton communities to clay and phosphate loading in a shallow reservoir. J Phycol 27:373-384.
Burkholder JM, Glasgow HB, Jr.
1997 *Pfiesteria piscicida* and other *Pfiesteria*-like dinoflagellates: Behavior, impacts, and environmental controls. Limnol Oceanogr 42:1052-1075.
Burkholder JM, Noga EJ, Hobbs CW, Glasgow HB Jr, Smith SA.
1992 New "phantom" dinoflagellate is the causative agent of major estuarine fish kills. Nature 358:407-410, Nature 360:768.
Cachon J, Cachon M.
1987 Parasitic dinoflagellates. In: Taylor FJR, ed. The Biology of Dinoflagellates. Botanical Monographs. Vol. 21. Boston: Blackwell Scientific Publications. p 571-610.
Carmichael WW.
1994 The toxins of cyanobacteria. Sci Am 270:78-86.
Cembella AD, Milenkovic L, Doucette G, Fernandez M.
1995 In vitro biochemical methods and mammalian bioassays for phycotoxins. Part A. In vitro biochemical and cellular assays. In: Hallegraeff GM, Anderson DM, Cembella AD, eds. Manual on Harmful Marine Microalgae. Intergovernmental Oceanographic Commission Manuals and Guides No. 33. Paris: United Nations Educational, Scientific and Cultural Organization. p 177-211.
Chorus I, Bartram J, eds.
1999 Toxic Cyanobacteria in Water—A Guide to Their Public Health Consequences, Monitoring and Management. New York: E & FN Spon, on behalf of the World Health Organization.
Edvardsen B, Paasche E.
1998 Bloom dynamics and physiology of *Prymnesium* and *Chrysochromulina*. In: Anderson DM, Cembella AD, Hallegraeff GM, eds. Physiological Ecology of Harmful Algal Blooms. NATO ASI Series G. Ecological Sciences. Vol. 41. New York: Springer-Verlag. p 193-208.
Epstein PR.
1998 Marine Ecosystems: Emerging Diseases as Indicators of Change. Year of the Ocean Special Report. Boston: Center for Health and the Global Environment, Harvard Medical School.
Estep KW, MacIntyre F.
1989 Taxonomy, life cycle, distribution and dasmotrophy of *Chrysochromulina*: A theory accounting for scales, haptonema, muciferous bodies and toxicity. Mar Ecol Prog Ser 57:11-21.
Fairey ER, Edmunds JS, Deamer-Melia NJ, Glasgow HB Jr, Johnson FM, Moeller PR, Burkholder JM, Ramsdell JS.
1999 Reporter gene assay for fish killing activity produced by *Pfiesteria piscicida*. Environ Health Perspect 107:711-714.
Falconer I, ed.
1993a Algal Toxins in Seafood and Drinking Water. New York: Academic Press.
Falconer I.
1993b Measurement of toxins from blue-green algae in water and foodstuffs. In: Falconer I, ed. Algal Toxins in Seafood and Drinking Water. New York: Academic Press. p 165-175.

Faust MA.
1992 Observations on the morphology and sexual reproduction of *Coolia monotis* (Dinophyceae). J Phycol 28:94-104.
Faust MA.
1993 Sexuality in a toxic dinoflagellate, *Prorocentrum lima*. In: Smayda TJ, Shimizu Y, eds. Toxic Phytoplankton in the Sea. Amsterdam: Elsevier Science Publishers. p 121-126.
Franks PJS.
1995 Sampling techniques and strategies for coastal phytoplankton blooms. In: Hallegraeff GM, Anderson DM, Cembella AD, eds. Manual on Harmful Marine Microalgae. Intergovernmental Oceanographic Commission Manuals and Guides No. 33. Paris: United Nations Educational, Scientific and Cultural Organization. p 25-43.
Gallagher JC.
1998 Genetic variation in harmful algal bloom species: An evolutionary ecology approach. In: Anderson DM, Cembella AD, Hallegraeff GM, eds. Physiological Ecology of Harmful Algal Blooms. NATO ASI Series G. Ecological Sciences. Vol. 41. New York: Springer-Verlag. p 225-242.
Gentien P, Arzul G.
1990 Exotoxin production by *Gyrodinium* cf. *aureolum* (Dinophyceae). J Mar Biol Assn UK 70:571-581.
Glasgow HB Jr, Burkholder JM, Schmechel DE, Tester PA, Rublee PA.
1995 Insidious effects of a toxic dinoflagellate on fish survival and human health. J Toxicol Environ Health 46:501-522.
Gorham PR, Carmichael WW.
1988 Hazards of freshwater blue-green algae. In: Lembi CA, Waaland JR, eds. Algae and Human Affairs. New York: Cambridge University Press. p 403-431.
Grattan LM, Oldach D, Perl TM, Lowitt MH, Matuszak DL, Dickson C, Parrott C, Shoemacher RC, Wasserman MP, Hebel JR, Charache P, Morris JG Jr.
1998 Problems in learning and memory occur in persons with environmental exposure to waterways containing toxin-producing *Pfiesteria* or *Pfiesteria*-like dinoflagellates. Lancet 352:532-539.
Hallegraeff GM.
1993 A review of harmful algal blooms and their apparent global increase. Phycologia 32:79-99.
Hallegraeff GM, Anderson DM, Cembella AD.
1995 Manual on Harmful Marine Microalgae. Intergovernmental Oceanographic Commission Manuals and Guides No. 33. Paris: United Nations Educational, Scientific and Cultural Organization.
Hansen PJ.
1998 Phagotrophic mechanisms and prey selection in mixotrophic phytoflagellates. In: Anderson DM, Cembella AD, Hallegraeff GM, eds. Physiological Ecology of Harmful Algal Blooms. NATOASI Series G. Ecological Sciences. Vol. 41. New York: Springer-Verlag. p 525-538.
Haystead TJ, Sim ATR, Carling D, Honner RC, Tsukitani Y, Cohen P, Hardie DH.
1989 Effects of the tumour promotor okadaic acid on intracellular protein phosphorylation and metabolism. Nature 337:78-81.
Hokama Y, Miyahara JT.
1986 Ciguatera poisoning: Clinical and immunological aspects. J Toxicol Toxin Rev 5:25-62.
Howell BR, Shelton RGJ.
1970 The effect of China clay on the bottom fauna of St. Austell and Mevagissey Bays. J Mar Biol Assn UK 50:593-607.

Jacobson DM, Anderson DM.
1996 Widespread phagocytosis of ciliates and other protists by marine mixotrophic and heterotrophic thecate dinoflagellates. J Phycol 32:279-285.
Johannessen JA, Johannessen OM, Haugen PM.
1989 Remote sensing and model simulation studies of the Norwegian Coastal Current during the algal bloom in May 1988. Int J Remote Sensing 10:1893-1906.
Kirk JTO.
1994 Light and Photosynthesis in Aquatic Ecosystems, 2nd ed. New York: Cambridge University Press.
Landsberg JH.
1996 Neoplasia and biotoxins in bivalves: Is there a connection? J Shellfish Res 15:203-230.
Landsberg JH, Steidinger KA.
1998 A historical review of *Gymnodinium breve* red tides implicated in mass mortalities of the manatee (*Trichechus manatus* latirostris) in Florida, USA. In: Reguera B, Blanco J, Fernández ML, Reguera B, Blanco J, Fernández ML, and Wyatt T, eds. Harmful Algae. Santiago: VIII International Conference on Harmful Algae. GRAFISANT. p 97-100.
Lembi CA, O'Neal SW, Spencer DF.
1988 Algae as weeds: Economic impact, ecology, and management alternatives. In: Lembi CA, Waaland JR, eds. Algae and Human Affairs. New York: Cambridge University Press. p 455-481.
Levin ED, Simon BB, Schmechel DE, Glasgow HB Jr, Deamer-Melia NJ, Burkholder JM, Lewis RJ.
1995 Detection of ciguatoxins and related benthic dinoflagellate toxins: In vivo and in vitro methods. In: Hallegraeff GM, Anderson DM, Cembella AD, eds. Manual on Harmful Marine Microalgae. Intergovernmental Oceanographic Commission Manuals and Guides No. 33. Paris: United Nations Educational, Scientific and Cultural Organization. p 135-161.
Lewitus AJ, Glasgow HB Jr, Burkholder JM.
1999 Kleptoplastidy in the toxic dinoflagellate, *Pfiesteria piscicida*. J Phycol 35:303-312.
MacKenzie L.
1992 Does *Dinophysis* (Dinophyceae) have a sexual life cycle? J Phycol 28:399-406.
Magnien RE.
2000 The dynamic of science and policy during the outbreak of *Pfiesteria* in Chesapeake Bay. BioScience (Forthcoming).
Milligan KLD, Cosper EM.
1994 Isolation of virus capable of lysing the brown tide microalga, *Aureococcus anophagefferens*. Science 266:805-807.
Moser VC, Jensen K, Harry GT.
1999 *Pfiesteria* toxin and learning performance. Neurotoxicol Teratol 21:215-221.
Nakajima I, Oshima Y, Yasumoto T.
1981 Toxicity of benthic dinoflagellates in Okinawa. Bull Jpn Soc Fishery Sci 47:1029-1033.
Oldach DW, Delwiche CF, Jakobsen KS, Torstein T, Brown EG, Kempton JW, Schaefer EF, Bowers H, Glasgow HB Jr, Burkholder JM, Steidinger KA, Rublee PA.
2000 Heteroduplex Mobility Assay guided sequence discovery: Elucidation of the small subunit (18S) rDNA sequence of *Pfiesteria piscicida* from complex algal culture and environmental sample DNA pools. Proc Natl Acad Sci 97:4304-4308.
Peleaz J.
1987 Satellite images of a "red tide" episode off southern California. Oceanol Acta 10:403-410.

Pérez E, Martin DF.
1999 Mitigation of harmful algal blooms, an annotated bibliography: 1986-1998. Environ Chem 62:115-127.
Pfiester LA, Popovsky J.
1978 Parasitic amoeboid dinoflagellates. Nature 379:421-424.
Popovský J, Pfiester LA.
1990 Süßwasserflora von Mitteleuropa—*Dinophyceae* (*Dinoflagellida*). Stuttgart: Gustav Fischer Verlag.
Ross MA, Lembi CA.
1999 Applied Weed Science, 2nd ed. Saddle River, NJ: Prentice Hall.
Russell FE, Egen NE.
1991 Ciguateric fishes, ciguatoxin (CTX) and ciguatera poisoning. J Toxicol Toxin Rev 10:37-62.
Schnepf E, Elbrächter M.
1992 Nutritional strategies in dinoflagellates—A review with emphasis on cell biological aspects. Eur J Protistol 28:3-24.
Scholin CA.
1998a Development of nucleic acid probe-based diagnostics for identifying and enumerating harmful algal bloom species. In: Anderson DM, Cembella AD, Hallegraeff GM, eds. Physiological Ecology of Harmful Algal Blooms. NATO ASI Series G. Ecological Sciences. Vol. 41. New York: Springer-Verlag. p 337-349.
Scholin CA.
1998b Morphological, genetic and biogeographic relationships of toxic dinoflagellates *Alexandrium tamarense*, *A. catenella* and *A. fundyense*. In: Anderson DM, Cembella AD, Hallegraeff GM, eds. Physiological Ecology of Harmful Algal Blooms. NATO ASI Series G. Ecological Sciences. Vol. 41. New York: Springer-Verlag. p 13-28.
Scholin CA, Gulland F, Doucette GJ, Benson S, Busman M, Chavez FP, Cordaro J, DeLong R, DeVogelmere A, Harvey J, Haulena M, Lefebvre K, Lipscomb T, Loscutoff S, Lowenstine LJ, Martin R, Miller P, McLellan WA, Moeller PDR, Powell CL, Rowles T, Silvagni P, Silver M, Spraker T, Trainer V, Van Dolah FM.
2000 Mortality of sea lions along the central California coast linked to a toxic diatom bloom. Nature 403:80-84.
Shumway SE.
1995 Phycotoxin-related shellfish poisoning: bivalve molluscs are not the only vectors. Rev Fishery Sci 3:1-31.
Skulberg OM, Carmichael WW, Codd GA, Skulberg R.
1993 Taxonomy of *Cyanophyceae* (cyanobacteria). In: Falconer I, ed. Algal Toxins in Seafood and Drinking Water. New York: Academic Press. p 145-164.
Sperr AE, Doucette GJ.
1996 Variation in growth rate and ciguatera toxin production among geographically distinct isolates of *Gambierdiscus toxicus*. In: Yasumoto T, Oshima T, Fukuyo Y, eds. Harmful and Toxic Algal Blooms. Paris: Intergovernmental Oceanographic Committee of UNESCO. p 309-312.
Steidinger KA.
1983 A re-evaluation of toxic dinoflagellate biology and ecology. In: Round FE, Chapman DJ, eds. Progress in Phycological Research. Vol. 2. New York: Elsevier. p 147-188.
Steidinger KA, Vargo GA, Tester PA, Tomas CR.
1998 Bloom dynamics and physiology of *Gymnodinium breve* with emphasis on the Gulf of Mexico. In: Anderson DM, Cembella AD, Hallegraeff GM, eds. Physiological Ecology of Harmful Algal Blooms. NATO ASI Series G. Ecological Sciences. Vol. 41. New York: Springer-Verlag. p 133-154.

Steidinger KA, Burkholder JM, Glasgow HB Jr, Hobbs CW, Truby E, Garrett J, Noga EJ, Smith SA.
  1996 *Pfiesteria piscicida* gen. et sp. nov. (*Pfiesteriaceae*, fam. nov.), a new toxic dinoflagellate genus and species with a complex life cycle and behavior. J Phycol 32:157-164.
State of Maryland.
  1998 Water Quality Improvement Act of 1998. Annapolis, MD: Maryland General Assembly.
Taylor FJR.
  1987 Dinoflagellate morphology. In: Taylor FJR, ed. The Biology of Dinoflagellates. Botanical Monographs. Vol 21. Boston: Blackwell Scientific Publications. p 24-92.
Taylor FJR, Pollingher U.
  1987 Ecology of dinoflagellates. A. General and marine ecosystems. In: Taylor FJR, ed. The Biology of Dinoflagellates. Botanical Monographs. Vol 21. Boston: Blackwell Scientific Publications. p 398-502.
Walker LM, Steidinger KA.
  1979 Sexual reproduction in the toxic dinoflagellate *Gonyaulax monilata*. J Phycol 15: 312-315.
Wetzel RG.
  1983 Limnology. Philadelphia: Saunders.
WHOI [Woods Hole Oceanographic Institute].
  2000 Glossary of *Pfiesteria*-Related Terms. *Pfiesteria* Interagency Coordination Work Group, chaired by J. Macknis, US EPA, Baltimore, MD. Available at <http://www.redtide.whoi.edu/pfiesteria/documents/glossary.html>.
Yu Z, Zou J, Ma X.
  1995 Application clays for removal of red tide organisms. III. The coagulation of kaolin on red tide organisms. Chinese J Oceanol Limnol 13: 62-70.

The Need for New Biotechnological Tools for Conservation of Marine Environments

Michael Smolen

World Wildlife Fund is a conservation organization. We do not have laboratories; however, we have active science programs in which we seek answers to questions we think are critical for responsible management of our natural resources. We work with scientists because we have learned that by working with scientists, we can answer questions that each of us alone cannot answer. Such questions may be "What are the impacts of certain types of perturbations on community structure?" or "How can we preserve for future generations critical species, communities, and habitats?" Answers to questions such as these require partnerships with many scientists in many different disciplines.

The view I am going to give you today is the view of a person who is not in the laboratory, and who is not currently involved with a particular application of marine biotechnology. Nevertheless, I will reinforce the voices of many who spoke here who call for more basic research. I will describe a need for new biotechnologies to (1) better understand community structure and changes among its components; (2) study possible linkages of disease and die-offs with changes in microorganisms in the marine community; (3) detect perturbations to marine systems from anthropogenic activities; (4) study immune suppression in marine species; and (5) develop new genetic methodologies to establish origins and track the spread of exotic species.

Wildlife and Contaminants Program, World Wildlife Fund, Washington, DC

There has been much discussion today of remediation of oil spills, which is a single point source problem. However, dolphin deaths, coral die-offs, and eutrophication over a wide area are much more complex. Understanding the response of marine mammals, fish, and coral requires a better understanding of the normal processes and community structure to isolate and identify the causative agents. Most issues of environmental and conservation concern today are not as starkly simple as spilled oil.

We lack the knowledge to understand the linkages from the lower scale of organization (e.g., viruses, bacteria, protozoans) and the dynamics that occur at this level. Finding linkages with changes at the lower scale will help identify effects higher up the scale, which might explain events like the marine mammal die-offs. There need to be discussions on the role of morbillivirus, red tide, or possibly other organisms in dolphin, manatee, and fish deaths. As many of you have stated here today, we also need a better understanding of the basic biology and processes in marine systems. Filling data gaps requires new biotechnology to collect and new ways to analyze these data.

There is also a serious need for new biotechnologies that will yield new data and insights about community structure. Measurements that are cheaper, simpler, faster, and easier are needed. Applications that assess wider scales should provide a better picture of the events that actually occur over a wide geographic area, and new biotechnologies should better identify the causes of perturbations. It is important to identify these perturbations early to minimize the effects on communities and economies.

Conservation questions involve both natural species and human societies living along the coasts of marine systems. Potential conflicts exist between human activities and processes in the natural marine systems. One area of concern is coastal mariculture. As a conservation organization, we must ask about the consequences of growing shellfish in a confined area:

- Because shellfish release chemical messages that coordinate and regulate reproduction at the population level, what effects do these high-density populations have on other naturally occurring species?
- Are they affecting the growth, reproduction, and population sizes of the natural populations?

Marine aquaculture, such as salmon rearing, is another form of high-density farming along the coastlines.

- What are the effects on the marine communities from this style of farming and the nutrient enrichment associated with the feeding regimes?
- Are the natural communities in the bays, estuaries, and coastlines changing because of these anthropogenic activities?

These questions again point out the need for more basic research and the new biotechnological tools necessary to detect changes to community structure and function. Without such insights, we increase the risk of permanent wide-scale changes that will necessitate broad regulatory action.

We need to better understand the immune system of fishes, mammals, and coral:

- Are the die-offs that we are seeing with greater frequency caused or enhanced because of immune suppression?

There is a growing list of chemicals with varying degrees of biomagnification and persistence that can affect various components of the immune system. The competency of the immune systems of these species must be studied to determine whether changes in exposure increase their susceptibility to disease or parasites normally resident in the marine environment.

- Does immune suppression afford these pathogens the opportunity to have a greater impact?

The speakers here today stressed that such questions cannot be answered without more basic information about the species. An additional focus on the new methodologies and tools is necessary to quantify changes in the immune system and to correlate this with the effects of perturbation.

There is also a need to increase our understanding of marine community health and how we conserve community structure.

- Is there sufficient marine biotechnology to date that identifies all the primary producers?
- Can we readily identify changes occurring within the microorganisms? What are the potential effects, the projected effects, of perturbations to photosynthetic rates in phytoplankton?
- What does it mean to community structure?
- Is it a temporary effect or is it a persistent long-term effect?

More attention must be given to the separation of effects of chemicals, UVB, habitat change, or fisheries management practices. Greater attention should be given to documenting effects through biomarkers. We must move beyond recording the presence or absence of agents, beyond recording simple concentrations of synthetic chemicals or patterns of distribution. Biotechnology must be developed to assist in linking the presence of such agents to effects on species and communities.

New biotechnology is needed for better genetic identification. Exotic species are a growing threat, and we must seek a better and quicker method to identify them and determine their origins. As they come in on ships to our harbors and waterways, we must identify them quickly, simply, and effectively to prevent them from altering our natural communities and local economies. Quick identification can lead to better management plans and strategic regulations. There are agencies that can implement such technologies and craft them into programs of vigilance; however, they do not have the means or budget to develop these new technologies.

Lastly, there have been a number of appeals by those presenting today for increased educational outreach. There is also a need for more professional educational outreach, for new partnerships between the basic and applied scientists and for partnerships with conservation organizations. These three sectors have their particular strengths that must be shared if the marine coastal regions are to be fully understood and protected. Professional outreach is necessary to transfer the hard science into sound management practices. Likewise, the changes in the coastal communities must help to redirect science to ask new questions. This is best continued through basic science and regular discussions.

Social and Regulatory Aspects of the Marine Environment

Raymond A. Zilinskas

There appears to be a consensus among workshop presenters on the need to widely publicize two subjects pertaining to marine biotechnology: the problems that can be alleviated or solved by biotechnological applications and the success stories of biotechnological applications. I hasten to add that we cannot and should not assume the vast burden of publicizing what biotechnology has done and might do; however, it is important to make known more widely the accomplishments and promises of *marine* biotechnology.

I was shocked earlier when I heard about severely damaged corals and reefs. Because I am a SCUBA diver and have visited reefs in many parts of the world, I think of myself as somebody who should be knowledgeable about these problems. Yet until yesterday, I had not known of the awful disease and other threats to the health of corals and reefs. I would estimate that the level of ignorance among the general public is at least equal to mine. So how do we inform the public about these threats that well might endanger the well-being of our planet? Furthermore, what can we do about presenting to the public information about how biotechnological solutions might be applied to counter these threats?

Monterey Institute of International Studies, Monterey, CA

SOCIAL ASPECTS

Identifying Public Perception

I know from personal experience that educating the public about scientific and technical advances can be very difficult. In 1996 and 1997, I managed a project called "The Human Genome Project: Reaching the Minority Communities in Maryland." As the name suggests, the objective of the project was to inform minority communities in Maryland about the Human Genome Project (HGP) as well as the ethical, legal, and social issues it might engender (Zilinskas and Balint 2000). The method decided on to convey this information was to hold a 2-day conference to which approximately 200 community leaders from mainly the Hispanic and African-American communities were invited. The first morning of the conference was devoted to presenting basic facts about the HGP. I will never forget when, during the discussion that followed that first morning, a lady stood up and asked, "Why have they been keeping this information from us? Who are they that keep this information from us?" By the time the conference ended, it had been demonstrated to the participants that nobody was suppressing information about the HGP; that there had been numerous articles in newspapers and magazines, as well as reporting on television and radio on the subject. Yet at the beginning of the conference, there appeared to have been a general perception among participants that important information had been withheld from them.

Afterward, I thought about possible reasons why this misperception of information being withheld would appear. One reason might be that the information is being presented in such a way that it is not readily understood by laypersons. Another reason could be that for each one of us there is a kind of self-censorship at work when we are presented with a mass of information. I know that when I read newspapers, magazines, and journals and when I watch television, I select reading or watching subjects that are of most interest to me. If I have time, I might move to subjects of potential or secondary interest. Therefore, because I find the HGP and its social implications interesting, I will read articles or watch programs that deal with these topics. However, people who have not been introduced to these topics probably would focus on subjects of more interest to them and skip right by anything to do with the HGP.

In any case, for almost all of the conference participants, the HGP was an unknown subject, as demonstrated by results from a survey of the participants before, after, and 1 year after the conference. The survey also demonstrated that once the participants had a good introduction to the subject of the HGP, it remained interesting to them and they tended to follow developments related to it. In fact, some of them got involved in the political process at the state level in legislative activities that aimed to address privacy and other concerns.

What I try to show by telling this story is that we have to find a way to make marine biotechnology understandable to more members of the public and thus get them interested in following developments related to this area of science. To do this, we must go beyond what, for example, Sea Grant does. Maryland Sea Grant has a very nice little newsletter they send out quarterly; I think it is wonderful, and I use it as a basic resource. However, I do not think it is known to the general public who might be interested or should be aware of these things, especially at the state legislative or federal congressional level. Although I am not certain how we do it, I think it is very important to include this aspect of how to deliver information about marine biotechnology effectively to the public and its representatives in the package of requests for proposals (RFPs) that the sponsors of this workshop might develop.

Measuring Prospective Support

We might also recommend something about testing or measuring public perceptions about marine biotechnology. As Dr. Prince stated, almost as a "truism," everyone accepts bioremediation. Although this may be true, I would like to learn whether this truism has a firm foundation. To do so, this workshop's sponsors might consider supporting a project that would measure public perception about a prospective bioremediation activity in an ecologically sensitive region such as the Chesapeake Bay, the Santa Barbara Channel, and the Puget Sound (see Mendelssohn, this volume). Any of these areas could at any time be the site of a serious oil spill from either shipwrecks or leaky off-shore installations. Such a project might involve setting up a focus group whose membership would consist of the major stakeholders—people who manage beaches, tourist promotion agencies, local fisheries, public health officials, and citizens from local communities. This focus group could be presented with a scenario of a major oil spill and be asked to consider different options of dealing with it, including bioremediation. A project such as this probably would, at a fairly low price, provide a good idea of how those most likely to be affected by an oil spill would want this kind of disaster managed, including determining how they would view bioremediation and its application. So, those are my major comments about public perception.

REGULATORY ASPECTS

Existing Regulations

We appear to have many regulations that govern activities in the marine environment. A short time ago I read an article about mariculture (i.e., aquaculture in the marine environment) in which it was stated that mariculture is the most regulated industrial activity in the United States, even more regulated than the nuclear power industry. Although I have no way to judge the accuracy of this statement (especially since someone who wanted more freedom for the aquaculture industry wrote it), there are in fact many regulations for putting anything new into the marine environment, possibly involving the federal government, regional boards and agencies and, at times, international agreements (Stenquist 1998).

What would be the reaction of regulatory agencies if somebody proposed to coat ships or man-made structures in the ocean with a newly developed proteinaceous substance that prevented biofouling? I do not think we know. The people here from industry do not seem to be worried about it. Is that because they have researched this subject and have found that there is nothing to worry about, or is it because they do not know and are not going to worry about it until they have a new product slated to be applied in the marine environment? If indeed regulatory problems might attend such an introduction, then such an advance in marine biotechnology might well be hindered.

Recommendations

I certainly do not favor more regulations. What would be helpful for researchers and developers involved in marine biotechnology would be to investigate whether existing regulations could, first, be simplified and, second, be put on a firm scientific basis. The major reason for attempting these two steps would be to create a stable and straightforward regulatory path for developers who would like to introduce a new application based on biotechnology into the marine environment. Therefore, I suggest that it would be timely for agencies interested in sponsoring scientific research in marine biotechnology to also sponsor social science research that would cast light on the regulatory barriers that might hinder introductions of materials into the seas.

I also suggest that agencies interested in sponsoring research in marine biotechnology make a special effort to encourage basic research that would be undertaken for the purpose of generating data and information to meet the four familiarity criteria that underlie the EPA's 21 Points to Consider as well as those of the USDA and the Organization for Economic Cooperation and Development (an intergovernmental organization headquartered in Paris to which most of the world's industrialized nations belong). This effort would most likely require a multidisciplinary effort that would involve marine ecologists, marine ichthyologists, marine microbiologists, oceanographers, risk assessors, and specialists in other disciplines. Such an effort is necessary if we are to move ahead with the development of genetically modified organisms (GMOs) for applications in the marine environment, including mariculture and bioremediation.

ANTICIPATING THE FUTURE

Already, between 25 and 35 fish species have been transformed (i.e., a transformed organism is one that has DNA from a foreign source incorporated in its genome), as well as an unknown number of shellfish and marine bacterial species. While researching developments in marine biotechnology in preparation for writing a recent book (Zilinskas and Balint 1998), I heard rumors about the introduction of genetically modified fish into the marine environment of some Asian nations (possibly China, Taiwan, and Thailand). Even if we were to disregard these rumors, it is a fact that much of this kind of scientific development is being undertaken by Asian nations, many of which depend on the seas to supply them with a substantial part of their food requirements. It is reasonable to believe that if scientists in one of these nations were to develop a transgenic fish species perceived as offering substantial benefits to mariculture, it would be applied for this purpose. If so, some would escape.

Even if we do not have the scientific knowledge to fulfill the familiarity criteria, it is safe to assume that the technology push will sooner or later result in the introduction of GMOs into the open marine environment. So instead of saying, "No, no, it is not going to happen, nobody is going to do it without proper preparation, we are not going to allow it to happen," we should assume that it will happen; perhaps not in the United States or other Western nations, but somewhere. If so, we might as well try to prepare for such an event. A good way to prepare is to try to generate scientific data about the marine environment, phenomena, and organisms that would make it possible for scientists and regulators to perform risk assessments of proposed introductions of GMOs into the oceans. Appropriate regulations could then be developed and adopted by national legislative bodies. All nations might not do so, but at least there would be a possibility for them doing so—something that is not the case today. For these reasons, I believe this would be an excellent program to undertake within the framework of what we are trying to do here.

Learning from the Past

It might be that useful lessons can be learned from past introductions of exotic marine microorganisms; in other words, when microorganisms occurring naturally in one part of the world were transported and introduced into a new environment. If we could learn the fate of such microorganisms and their effects, if any, on the environment into which they were introduced, we could begin to generate information that bears on the familiarity criteria. I know that James T. Carlton of Williams College has done important work on the environmental effects of introduced exotic macroorganisms (Carlton 1995; Cohen and Carlton 1998). However, it appears that no one has done substantial work on the environmental effects of introduced exotic marine microorganisms. Of course, such investigations would be exceedingly difficult to carry out, but I do think it is possible with the help of the new biotechnologies and the recently developed tools of molecular biology such as polymerase chain reaction.

It is reasonable to hypothesize that similar to introductions of exotic microorganisms in the terrestrial environment, exotic marine microorganisms introduced into a new site could have beneficial, innocuous, or negative effects on that site's environment. If an organism were introduced deliberately, the assumed intent would be to achieve a benefit. However, the opposite possibility cannot be ignored. An illustrative example has been reported in a study by the Joint Subcommittee on Aquaculture Shrimp Virus Work Group (1997), which discussed the aftermath of the transportation of shrimp seed stocks from Asia to Central America and the United States. The imported seed stocks did improve the stocks of shrimp being aquacultured in the importing countries; however, the imports were contaminated with pathogenic viruses. As a result of these importations, some of the diseases that had devastated the shrimp aquaculture industry in several Asian nations were imported for the first time to Central America and the United States, where they caused heavy damage.

International Awareness

I understand from Dr. Denno's talk yesterday that the United States is not a party to the Law of the Sea or the Convention on Biological Diversity, international agreements negotiated during the time of the Reagan Administration. Nevertheless, we must consider that perhaps they represent an international norm of research and industrial activity. If so, it makes sense for our researchers and industrialists to adhere to their strictures, probably with the expectation that eventually the United States will join these treaties. It then follows that there is a need for social science research to explore how the Law of the Sea and the Convention on Biological Diversity will affect, for example, existing US laws and regulations, activities by US industries in the marine environment, and investigations by US researchers in the oceans.

Perhaps the book Dr. Denno mentioned yesterday (Cicin-Sain and others 2000) will address some of these issues. But even if it does, it is safe to assume that all important issues will not be covered and that some of the issues it addresses will have to be investigated further. I think it is important for the agencies that wish to support marine biotechnology to fund research that explores linkages between the US regulatory regime and the international regime with the objective of determining how the two can be reconciled in such a way that minimum inconvenience will be experienced by our researchers and industrialists.

REFERENCES

Carlton JT.
 1995 Marine invasions and the preservation of coastal diversity. Endangered Species Update 12(4/5):1-3.

Cicin-Sain B, Knecht RW, Jang D, eds.
 2000 Policy Issues in the Development of Marine Biotechnology. Newark, DE: University of Delaware Center for the Study of Marine Policy. (Forthcoming).

Cohen AN, Carlton JT.
 1998 Accelerating invasion rate in a highly invaded estuary. Science 279:555-558.

Joint Subcommittee on Aquaculture Shrimp Virus Work Group.
 1997 An Evaluation of Potential Shrimp Virus Impacts on Cultured Shrimp and Wild Shrimp Populations in the Gulf of Mexico and Southeastern US Atlantic Coastal Waters. Washington, DC: National Marine Fisheries Service; US Department of Commerce; Animal and Plant Health Inspection Service; US Department of Agriculture; National Center for Environmental Assessment; US Environmental Protection Agency; and Fish and Wildlife Service, US Department of Interior.

Stenquist S.
 1998 Federal and state regulations relevant to uncontained applications of genetically engineered marine organisms. In: Zilinskas RA, Balint PJ, eds. Genetically Engineered Marine Organisms: Environmental and Economic Risks and Benefits. Amsterdam: Kluwer Academic Publishers. p 139-180.

Zilinskas RA, Balint PJ, eds.
 1998 Genetically Engineered Marine Organisms: Environmental and Economic Risks and Benefits. Amsterdam: Kluwer Academic Publishers.

Zilinskas RA, Balint PJ, eds.
 2000 The Human Genome Project and Minority Communities: Ethical, Social, and Political Dilemmas. Westport CT: Praeger Publishers.

Rapporteur Comments on the Bioremediation Session

Roger C. Prince

We have heard that bioremediation is an important and ethical approach to many environmental problems. Perhaps the best thing is that if successful, it is a permanent solution to the environmental problem. Environmental contaminants are of particular concern when they are bioavailable and are doing something to the environment. Almost by definition, bioremediation is likely to remedy this; successful bioremediation removes the biologically available material. Most of the competing technologies are not as final as this because they usually only move the problem. They may concentrate or reuse it, but the typical response is to pick it up and put it somewhere else. Even the more rigorous physical approaches such as thermal desorption and washing do not focus on the bioavailable material, and reducing the total contamination may not be as effective as bioremediation at removing this material. Bioremediation has the advantage that when the microbes have done what they can do to organic compounds, the organic compounds are usually essentially completely eliminated. That is not always true, but at least it is the major process that goes on in the bioremediation of organic compounds. Bioremediation is also relatively inexpensive, which means that the people who have to do it rather like it, and it does have at least some public support as an environmentally appropriate technology.

Corporate Research Laboratory, Exxon/Mobil Research & Engineering Co., Annandale, NJ

The US Environmental Protection Agency (EPA) has greatly supported bioremediation and has pushed its use in a variety of situations. Exxon probably would not have been allowed to use bioremediation in Alaska (Prince and Bragg 1997) without the EPA pushing strongly for it to be tried (EPA 1989), and I think their efforts have led to general public support for the technology.

When we think about marine spills, it is important to recognize that the US coastline is divided into jurisdictions with statutory groups that decide how they would handle spills (http://www.nrt.org). Bioremediation is typically included as a final polishing step for open shorelines, and it is unlikely to be the "frontline" approach except in remote locations. Most shorelines are too publicly essential for something as slow as current bioremediation to work.

Bioremediation is much more likely to be useful in places where time is not absolutely of the essence and where other processes such as physical removal of the oil with bulldozers are very difficult or too dangerous for work crews. It is also important that any clean-up strategy have a net environmental benefit (Baker 1995), which bioremediation can readily achieve because it is so noninvasive.

As we have heard from all of the speakers, there are many good reasons that bioremediation is valuable, both on land (NRC 1993; Prince 1998) and in the marine environment (Lee and deMora 1999; Lin and others 1999; Prince and others 1997, 1999b; Swannell and others 1999). A major issue for those of us who are, as it were, practitioners is to continue successful applications when necessary. One of the biggest issues we face is maintaining both public and regulatory support, and a major issue is that bioremediation tends to be slow. There is thus a real need to give responders and the general public some confidence that a bioremediation strategy is having the expected results. We need to be able to show that the approach is encouraging a real biological process that will lead to biodegradation and removal of the contaminant. There is a pressing need for interim measures of success, and several of us are working on this issue. We are working on portable instrumentation and tool kits that allow us to monitor the success of fertilizer application, and the initial stimulation of microbial activity (Prince and others 1999a), but there is an obvious opportunity for developing sensitive molecular probes that will assess microbial responses directly.

Dr. Lee addressed toxicity endpoints. We are somewhat saddled in the terrestrial environment with clean-up standards that set some sort of concentration level, typically a goal of x parts per million of a particular contaminant. An alternative, and perhaps more meaningful, endpoint would be a goal of some minimal level of toxicity in two or three appropriate routine tests (Mueller and others 1999; Potter and others 1999; Saterbak and others 1999). The EPA is also addressing monitoring the disappearance of genotoxicity with successful bioremediation at terrestrial sites (Brooks and others 1998; Hughes and others 1998), and there are obvious potential extensions of this work to marine sediments (Ho and others 1999). The US Geological Survey (USGS) is developing a toxicity identification evaluation protocol for sediments (Lebo and others 1999), with the goal of using it for monitoring remediation. Thus, although there is some work in this area, there is a real need for more research on this issue. One new approach is to use semipermeable membrane devices loaded with oils that mimic fish tissue; these can be exposed at the contaminated site, and subsequently analyzed for contaminants, in a more reproducible way than exposing living animals on-site (Macrae and Hall 1998; Parrott and others 1999; Utvik and Johnsen 1999). Developing modern molecular genetic tools for assaying toxicity may also revolutionize this area and have profound influences on how and when remediation activities should be conducted.

There is also a very pressing need to deal with the mixed contaminants found in dredged materials from harbors and estuaries (NRC 1997). Dr. Young described anaerobic processes that might target such contaminants. A biological technology for cleaning contaminated sediments would be very useful, but it must accommodate the fact that the anaerobic conditions that are slowing down the degradation of some contaminants (e.g., hydrocarbons) are also immobilizing others (e.g., metals). The complete degradation of extensively halogenated compounds such as polychlorinated biphenyls (Bedard and others 1998; Wu and others 1999) requires initial anaerobic reductive dehalogenation followed by aerobic degradation. If one made currently anaerobic-contaminated sediments aerobic, one might well speed up the degradation of some organic contaminants, but at the potential expense of mobilizing currently immobilized metals and slowing reductive degradation processes. The issue of handling such mixed contamination requires much more research, and modern molecular tools may have an important role to play once the fundamental microbiological processes are understood. The area of anaerobic degradation of organic pollutants is so new that it is quite likely that basic research in this area will open new avenues for bioprocessing.

There are, also, some surprises when considering the spectrum of applications of bioremediation. In Europe there is now concern over the large volumes of vegetable oils that are shipped by sea (Mudge 1995). On several beaches, for example, mats of vegetable oil have polymerized on the beach. One might have thought that vegetable oil would be readily biodegradable and a very easy target for bioremediation. However, under some circumstances, it is proving to be a long-lived contaminant. Thus, there are many areas in bioremediation as a response for marine spills where we are still being surprised.

Bioremediation is basically aiming to stimulate natural processes. Someone asked yesterday, "So, you mean if you just waited, it would happen anyway?" and the answer is essentially, "Yes." With our current knowledge of bioremediation, we are only speeding up the natural process; and if we are lucky, we stimulate it up to five-fold.

The other side of this issue is that there are some situations where the natural rate of biodegradation of a contaminant is fast enough that there is no pressing need to stimulate it. There is quite a bit of research in this area, because of course such an approach might be even cheaper than bioremediation. Relying on natural attenuation is being accepted as an appropriate response for some terrestrial spills (Chapelle 1999; Lahvis and others 1999; Lu and others 1999; McNab and Dooher 1998; Stapleton and Sayler 1998), usually with the proviso that the site be monitored to ensure that the contaminant is indeed degrading and not migrating. There are obvious opportunities for using modern molecular probes in studying and quantifying the phenomena associated with natural attenuation, and work for terrestrial applications is well under way (Stapleton and Sayler 1998).

As we heard in Dr. Portier's presentation, it is often more important to clean up the source of chronic contamination than to clean the contaminated site. If the point source is removed, natural attenuation may well remedy the contaminated area that was being affected by the source.

So, a part of what we heard was the need to continue and steadily improve the successful applications of bioremediation. However, another important avenue for research and development is to move bioremediation to the next level of effectiveness and speed. There is a general optimism in the field that we ought to be able to do radically better in stimulating natural processes without causing any significant harm—We need to find ways of getting things to happen much faster. In the laboratory, one can get many contaminants to disappear with dramatically rapid rate constants, but we are not within two orders of magnitude of these rates in the field. We have to understand better what it is that is limiting the biodegradation of some of our contaminants.

The presentations principally dealt with organic compounds, and the great thing about organic compounds is that eventually they are converted to CO_2 and water and to all intents and purposes disappear. Inorganic contaminants can only be moved or collected; and although there are several technologies for handling inorganic contaminants in wastewater (Krishnan and others 1993), including several biological ones (Keasling and others 1998; Kefala and others 1999), there are not yet any proven biotechnologies for dealing with metals and metalloids in sediments, shorelines, and marshes.

One potential new approach is phytoremediation—the use of plants to remedy environmental problems (Salt and others 1995). Although there is interest in using plants to stimulate oil biodegradation (Carman and others 1998), most work appears to be focused on using plants to accumulate metals and metalloids (Raskin and others 1994).

Thus, in summary, we heard that bioremediation in the marine environment is an important option for dealing with spills and a potential option for dealing with contaminated sediments. Progress in developing these technologies will come from a number of fronts, and modern molecular approaches must be integrated into this ongoing work. Terrestrial applications of bioremediation have received more attention than marine ones because of the far greater need, and gene-probe and molecular taxonomy approaches are beginning to move from the laboratory to the field. Some of these will be directly transferable to the marine environment, but there will undoubtedly be a need to develop saline-specific techniques. The next few years promise to be an exciting time for such developments.

REFERENCES

Baker JM.
 1995 Net environmental benefit analysis for oil spill response. In: Proceedings of the 1995 International Oil Spill Conference. Washington, DC: American Petroleum Institute. p 611-614.
Bedard DL, VanDort H, Deweerd KA.
 1998 Brominated biphenyls prime extensive microbial reductive dehalogenation of aroclor 1260 in Housatonic River sediment. Appl Environ Microbiol 64:1786-1795.
Brooks LR, Hughes TJ, Claxton LD, Austern B, Brenner R, Kremer F.
 1998 Bioassay-directed fractionation and chemical identification of mutagens in bioremediated soils. Environ Health Perspect 106(Suppl 6):1435-1440.
Carman EP, Crossman TL, Gatliff E.G.
 1998 Phytoremediation of no. 2 fuel oil-contaminated soil. J Soil Contam 7:455-466.
Chapelle FH.
 1999 Bioremediation of petroleum hydrocarbon-contaminated ground water: The perspectives of history and hydrology. Ground Water 37:122-132.
EPA [Environmental Protection Agency].
 1989 Alaskan oil spill bioremediation project. EPA/600/8-89/073.
Ho K, Patton L, Latimer JS, Pruell RJ, Pelletier M, McKinney R, Jayaraman S.
 1999 The chemistry and toxicity of sediment affected by oil from the North Cape spilled into Rhode Island Sound. Marine Poll Bull 38:314-323. (See also "The US National Response Team," <http://www.nrt.org>.)
Hughes TJ, Claxton LD, Brooks L, Warren S, Brenner R, Kremer F.
 1998 Genotoxicity of bioremediated soils from the Reilly tar site, St. Louis Park, Minnesota. Environ Health Perspect 106(Suppl 6):1427-1433.

Keasling JD, VanDien SJ, Pramanik J.
  1998   Engineering polyphosphate metabolism in *Escherichia coli*: Implications for bioremediation of inorganic contaminants. Biotechnol Bioengineer 58:231-239.
Kefala MI, Zouboulis AI, Matis KA.
  1999   Biosorption of cadmium ions by actinomycetes and separation by flotation. Environ Poll 104:283-293.
Krishnan ER, Utrecht PW, Patkar AN, Davis JS, Pour SG, Foerst ME.
  1993   Recovery of metals from sludges and wastewaters. Park Ridge, NJ. Noyes Data Corporation.
Lahvis MA, Baehr AL, Baker RJ.
  1999   Quantification of aerobic biodegradation and volatilization rates of gasoline hydrocarbons near the water table under natural attenuation conditions. Water Resources Res 35:753-765.
Lebo JA, Huckins JN, Petty JD, Ho KT.
  1999   Removal of organic contaminant toxicity from sediments—Early work toward development of a toxicity identification evaluation (TIE) method. Chemosphere 39:389-406.
Lee K, deMora S.
  1999   In situ bioremediation strategies for oiled shoreline environments. Environ Technol 20:783-794.
Lin, Q, Mendelssohn IA, Henry CB, Robert, PO, Walsh MM, Overton EB, Portier, RJ.
  1999   Effects of bioremediation agents on oil degradation in mineral and sandy salt marsh sediments. Environ Technol 20:825-837.
Lu GP, Clement TP, Zheng CM, Wiedemeier TH.
  1999   Natural attenuation of BTEX compounds: Model development and field-scale application. Ground Water 37:707-717.
Macrae JD, Hall KJ.
  1998   Comparison of methods used to determine the availability of polycyclic aromatic hydrocarbons in marine sediment. Environ Sci Technol 32:3809-3815.
McNab WW, Dooher BP.
  1998   Uncertainty analyses of fuel hydrocarbon biodegradation signatures in ground water by probabilistic modeling. Ground Water 36:691-698.
Mudge SM.
  1995   Deleterious effects from accidental spillages of vegetable oils. Spill Sci Technol Bull 2:187-191.
Mueller DC, Bonner JS, McDonald SJ, Autenrieth RL.
  1999   Acute toxicity of estuarine wetland sediments contaminated by petroleum. Environ Technol 20:875-882.
NRC [National Research Council].
  1997   Contaminated Sediments in Ports and Waterways: Cleanup Strategies and Technologies. Washington, DC. National Academy Press.
NRC [National Research Council].
  1993   In Situ Bioremediation: When Does It Work? Washington, DC: National Academy Press.
Parrott JL, Backus SM, Borgmann AI, Swyripa M.
  1999   The use of semipermeable membrane devices to concentrate chemicals in oil refinery effluent on the Mackenzie River. Arctic 52:125-138.
Potter CL, Glaser JA, Chang LW, Meier JR, Dosani MA, Herrmann RF.
  1999   Degradation of polynuclear aromatic hydrocarbons under bench-scale compost conditions. Environ Sci Technol 33:1717-1725.

Prince RC.
1998   Bioremediation. In: Kirk-Othmer Encyclopedia of Chemical Technology, 4th ed, suppl. New York: John Wiley. p 48-89.
Prince RC, Atlas RM, Zelibor JL Jr.
1997   Environmental applications of marine biotechnology. In: Altman A, ed. Agricultural Biotechnology. New York: Marcel Dekker. p 615-628.
Prince RC, Bare RE, Garrett RM, Grossman MJ, Haith CE, Keim LG, Lee K, Holtom G, Lambert P, Sergy GA, Owens EH, Guénette CC.
1999a Bioremediation of a marine oil spill in the Arctic. In: In Situ Bioremediation of Petroleum Hydrocarbon and Other Organic Compounds. Alleman BC, Leeson A, eds. Columbus, OH: Battelle Press. p 227-232.
Prince RC, Bragg JR.
1997   Shoreline bioremediation following the Exxon Valdez oil spill in Alaska. Biomed J 1:97-104.
Prince RC, Varadaraj R, Fiocco RJ, Lessard RR.
1999b Bioremediation as an oil spill response tool. Environ Technol 20:891-896.
Raskin I, Nanda Kumar PBA, Dushenkov S, Salt DE.
1994   Bioconcentration of heavy metals by plants. Curr Opinion Biotechnol 5:285-290.
Salt DE, Blaylock M, Nanda Kumar PBA, Dushenkov V, Ensley BD, Chet I, Raskin I.
1995   Phytoremediation: A novel strategy for the removal of toxic metals from the environment using plants. Biotechnology 13:468-474.
Saterbak A, Toy RJ, Wong DCL, McMain BJ, Williams MP, Dorn PB, Brzuzy LP, Chai EY, Salanitro JP.
1999   Ecotoxicological and analytical assessment of hydrocarbon-contaminated soils and application to ecological risk assessment. Environ Toxicol Chem 18:1591-1607.
Stapleton RD, Sayler GS.
1998   Assessment of the microbiological potential for the natural attenuation of petroleum hydrocarbons in a shallow aquifer system. Microbial Ecol 36:349-361.
Swannell RPJ, Mitchell D, Lethbridge G, Jones D, Heath D, Hagley M, Jones M, Petch S, Milne R, Croxford R, Lee K.
1999   A field demonstration of the efficacy of bioremediation to treat oiled shorelines following the Sea Empress incident. Environ Technol 20:863-873.
Utvik TIR, Johnsen S.
1999   Bioavailability of polycyclic aromatic hydrocarbons in the North Sea. Environ Sci Technol 33:1963-1969.
Wu QZ, Bedard DL, Wiegel J.
1999   2,6-Dibromobiphenyl primes extensive dechlorination of aroclor 1260 in contaminated sediment at 8-30 degrees C by stimulating growth of PCB-dehalogenating microorganisms. Environ Sci Technol 33:595-602.

Appendixes

Appendix A

WORKSHOP ON OPPORTUNITIES FOR ADVANCEMENT OF ENVIRONMENTAL APPLICATIONS OF MARINE BIOTECHNOLOGY

Georgetown Holiday Inn
2101 Wisconsin Avenue NW
Washington, DC 20007

OCTOBER 5-6, 1999

AGENDA

Tuesday, 5 October 1999
Introduction and Goals
8:30 a.m.   Roger C. Prince, Exxon/Mobil Research & Engineering
            Linda Kupfer, Sea Grant
            Maryanna Henkart, National Science Foundation

Biomaterials
9:00   *Introduction*—David Manyak, Oceanix Biosciences
9:10   — Marc W. Mittelman, Altan Corp.
9:30   — J. W. Costerton, Montana State University—Bozeman
        Antifouling
9:50   *Roundtable discussion*
10:20   Break
10:35   Rapporteur Report on Biomaterials—David Manyak

Economic and Regulatory Aspects
11:20   *Introduction* — Raymond A. Zilinskas, Monterey Institute
        of International Studies
11:30   — Lori Denno, Delaware Nature Society
        Regulatory considerations
11:50   — Diane Hite, Mississippi State University
        Economic considerations
12:10 p.m. Lunch
1:00   *Roundtable discussion*

Bioremediation

1:30	*Introduction* — Roger C. Prince, Exxon/Mobil Research & Engineering
1:40	— Lily Young, Rutgers University Spilled oil bioremediation
2:00	— Kenneth Lee, Department of Fisheries and Oceans, Canada Spilled oil bioremediation
2:20	— Ralph J. Portier, Louisiana State University Marsh bioremediation
2:40	— Irving A. Mendelssohn, Louisiana State University *Respondant*
2:50	*Roundtable discussion*
3:20	Break

Restoration

3:30	*Introduction* — Judith McDowell, Woods Hole Oceanographic Institution
3:40	— Aileen N. C. Morse, University of California—Santa Barbara Coral and reef restoration
4:00	— Laurie Richardson, Florida International University Coral epidemiology
4:20	— Richard Dodge, National Coral Reef Institute *Respondant*
4:30	*Roundtable discussion*
5:00	Adjourn

Wednesday, 6 October 1999

Restoration/Prediction and Monitoring

8:30 a.m.	*Introduction* — Roger C. Prince, Exxon/Mobil Research & Engineering
8:40	— François M. M. Morel, Princeton University Inorganic metals
9:00	— Jed Fuhrman, University of Southern California, Los Angeles Microbial contamination
9:20	— Mark E. Hahn, Woods Hole Oceanographic Institution Toxicology

9:40	— JoAnn Burkholder, North Carolina State University Algal Blooms
10:00	— Michael Smolen, World Wildlife Fund *Respondant*
10:10	*Roundtable discussion*
10:40	Break

Rapporteur Reports

11:00	— Raymond A. Zilinskas, Monterey Institute of International Studies Socioeconomic and regulatory aspects *General discussion*
11:45	— Roger C. Prince, Exxon/Mobil Research & Engineering Bioremediation *General discussion*
12:30 p.m.	Lunch
1:15	— Judith McDowell, Woods Hole Oceanographic Institution Restoration *General discussion*

Closing Comments

2:00	Roger C. Prince, Exxon/Mobil Research & Engineering
2:30	Adjourn

Appendix B

OPPORTUNITIES FOR ADVANCEMENT OF ENVIRONMENTAL MARINE BIOTECHNOLOGY

WORKSHOP PARTICIPANTS

JoAnn M. Burkholder, Department of Botany, North Carolina State University, Raleigh, NC
Linda Chrisey, Program Officer, Biomolecular and Biosystems Division, Office of Naval Research, Arlington, VA
Chrys Chryssostomidis, Massachusetts Institute of Technology, Cambridge, MA
J. W. Costerton, Center for Biofilm Engineering, Montana State University, Bozeman, MT
Lori Denno, Natural Resources Conservation, Delaware Nature Society, Hockessin, DE
Richard E. Dodge, Oceanographic Center, Nova Southeastern University, Dania Beach, FL
Jed Fuhrman, McCulloch-Crosby Chair of Marine Biology, University of Southern California, Los Angeles, CA
Mark E. Hahn, Biology Department, Woods Hole Oceanographic Institution, Woods Hole, MA
Maryanna Henkart, Division of Molecular & Cellular Biosciences, National Science Foundation, Arlington, VA
Rosemarie Hinkel, Center for the Study of Marine Policy, Graduate College of Marine Studies, University of Delaware, Newark, DE
Diane Hite, Department of Agricultural Economics, Mississippi State University, Mississippi State, MS
George Hoskin, DSATOS/OS/CFSAN, Food and Drug Administration, Washington, DC
Jonathan Kramer, Sea Grant, University Systems of Maryland, College Park, MD
Linda Kupfer, National Sea Grant College Program, OAR, National Oceanographic and Atmospheric Administration, US Department of Commerce, Silver Spring, MD

Kenneth Lee, Environmental Sciences Division, Maurice Lamontagne Institute, Fisheries and Oceans Canada, Mont-Joli, Quebec, Canada
Leonard Levin, Air Toxics Health and Risk Assessment, Electric Power Research Institute, Palo Alto, CA
David Manyak, Oceanix Biosciences, Hanover, MD
Judith McDowell, Woods Hole Oceanographic Institution, Woods Hole, MA
Irving A. Mendelssohn, Wetland Biogeochemistry Institute, and Department of Oceanography and Coastal Sciences, Louisiana State University, Baton Rouge, LA
Robert Menzer, US Environmental Protection Agency, Washington, DC
Marc W. Mittelman, Altran Corporation, Boston, MA
François M. M. Morel, Princeton Environmental Institute, Department of Geosciences, Princeton University, Princeton, NJ
Aileen N. C. Morse, Marine Biotechnology Center, Marine Science Institute, University of California, Santa Barbara, CA
Ralph J. Portier, Aquatic/Industrial Toxicology Laboratory, Institute for Environmental Studies, Louisiana State University, Baton Rouge, LA
Roger C. Prince, Corporate Research Laboratory, Exxon/Mobil Research & Engineering Co., Annandale, NJ
Laurie L.Richardson, Department of Biological Sciences, Florida International University, Miami, FL
Michael Smolen, Wildlife and Contaminants Program, World Wildlife Fund, Washington, DC
George Vermont, Division of Bioengineering and Environmental Systems, National Science Foundation, Arlington, VA
Cheryl Woodley, Charleston Laboratory, National Oceanic and Atmospheric Administration, Charleston, SC
Lily Young, Biotechnology Center for Agriculture and Environment, Rutgers University, New Brunswick, NJ
Raymond A. Zilinskas, Monterey Institute of International Studies, Monterey, CA